The Social Implications of Bioengineering

The Social Implications of Bioengineering

Elisabeth Beck-Gernsheim

Translated by
Laimdota Mazzarins

HUMANITIES PRESS
NEW JERSEY

Originally published in German as *Technik, Markt und Moral*, ©1991 by Fischer Taschenbuch Verlag GmbH, Frankfurt am Main.

This English edition first published 1995 by Humanities Press International, Inc., 165 First Avenue, Atlantic Highlands, New Jersey 07716.

English translation ©1995 by Humanities Press International, Inc.

Library of Congress Cataloging-in-Publication Data
Beck-Gernsheim, Elisabeth, 1946–
  [Technik, Markt und Moral. English]
  The social implications of bioengineering / Elisabeth Beck-Gernsheim : translated by Laimdota Mazzarins.
    p. cm.
  Includes bibliographical references and index.
  ISBN 0-391-03841-9 (cloth).—ISBN 0-391-03842-7 (paper)
  1. Human reproductive technology—Moral and ethical aspects.
  2. Genetic engineering—Moral and ethical aspects.
  I. Title.
  RG133.5.B4313 1995
  176—dc20   95-12912
                                    CIP

A CIP record for this book is available from the British Library.

Printed in the United States of America

Contents

I am not only convinced that what I say is wrong, but also that whatever will be said against it is wrong. Nonetheless, we must begin to talk about it. With a subject of this kind, the truth lies not in the middle but all around it, like a sack that changes its shape with every new opinion that is stuffed into it, yet becomes increasingly solid.

Robert Musil

Foreword to the English Edition

Reproductive medicine and genetic technology have become the objects of much discussion—in society and politics, in science and the media, and among the general public. The titles and headlines of many books and newspapers are signs of this trend. In the United States as well as in Germany, people are talking about the "Mysteries of Heredity"[1] and the "New Origins of Life,"[2] about "Happiness out of the Test Tube,"[3] and about "High-Tech Babies"[4] and "Baby Factories."[5] Elaborate promises are being made, and dire warnings are being voiced. The "Tailor-Made Human Being"[6] and "Programmed Heredity"[7] are becoming the vision of a possible future.

If we take a closer look at the debate, interesting differences become visible. In the United States a basic attitude generally dominates that is oriented toward the liberal market model and postulates freedom of choice; the discussion in Europe—especially in the Federal Republic of Germany and Switzerland—is clearly more restrictive in its basic orientation. A much larger number of groups is expressing its reservations here, and they are getting responses from the media and the general public. A great variety of reasons motivate these different groups to criticize: representatives of the churches see manipulation of God's creative work; supporters of the Greens, a rape of nature; representatives of the women's movement, a further attempt by the patriarchy to oppress women. But a certain reservation is becoming evident that cuts right through all of these groups and points toward the experiences of the German past. When the issue is genetic technology and its use on human beings, the memory of eugenics looms, together with the fatal, murderous consequences of the National Socialist policy, which drew a distinction between genetically "good" and genetically "inferior" human beings and carried out a selection process according to these criteria. The public discussion touches upon this experience of the entanglement of science in social issues and its political exploitation, which reached its pinnacle in the unscrupulous and blind obsession with research that became a "Lethal Science."[8] It is the awareness of this past

ix

that has sensitized people in areas of public concern, in legislation, science, politics and political parties, and even within the medical profession itself: "Among their international colleagues, the human geneticists of Germany are notorious on account of their anxiety" as far as the imputation of eugenics is concerned.[9]

What I have alluded to here in a few sentences can be observed in many concrete examples. Over and over again one can see cultural differences emerging clearly in the perception, evaluation, and assessment of the new biotechnologies. The cultural and historical background distracts our gaze, impels us to ask questions, and contributes to the hopes as well as the fears that we associate with this subject.

This also holds true—and how could it be otherwise?—for this book. On some points at least, it will sketch out a somewhat different perspective, a more "European" way of looking at reproductive medicine and genetic technology. The American reader may find new emphases and questions here. Many of these may seem quite stimulating and workable, although others, by contrast, will tend to irritate and provoke (and—who knows?—perhaps both kinds may even be quite useful in the end).

My hope is that an exchange of ideas and a dialogue with American colleagues who are working on similar subjects may perhaps develop, stimulated by this book. Is this an overly ambitious hope? Not necessarily. During the years since the German version of this book was published, an intense dialogue has developed between me and some scientists studying human genetics. We have learned to talk to one another and learn from one another outside the tribal rituals of the various scientific disciplines. The result is a new book that has been written by natural scientists and social scientists. It discusses, from a great variety of different perspectives, the question that unites us, moves us, motivates us: "Welche Gesundheit wollen wir?" (What Type of Health Do We Want?)[10]

Elisabeth Beck-Gernsheim

References

1. "Solving the Mysteries of Heredity," *Time*, 20 March 1989 (title article).
2. "The New Origins of Life," *Time*, 10 September 1984 (title article).
3. "Das Glück aus der Retorte" (Happiness Out of the Test Tube). *Stern*, no. 23/1984 (title article).

4. "High-Tech Babies." *Newsweek*, 18 March 1985 (title article).
5. "Baby-Fabriken" (Baby Factories). *Spiegel*, no. 17/1992 (title article).
6. Wolfgang van den Daele. *Mensch nach Maß? Ethische Probleme der Genmanipulation und Gentherapie (Tailor-Made Human Beings? Ethical Problems of Genetic Manipulation and Genetic Therapy)*. Beck: Munich, 1985.
7. Hans Harald Bräutigam and Liselotte Mettler. *Die programmierte Vererbung. Möglichkeiten und Gefahren der Gentechnologie* (Programmed Heredity: Possibilities and Dangers of Genetic Technology). Hamburg: Hoffmann und Campe, 1985.
8. Benno Müller-Hill, Tödliche Wissenschaft. *Die Aussonderung von Juden, Zigeunern und Geisteskranken 1933–1945* (Lethal Science: the Exclusion of Jews, Gypsies, and the Mentally Ill 1933–1945). Rowohlt: Reinbek, 1984.
9. *Spiegel*, no. 3/1993, p. 187.
10. Elisabeth Beck-Gernsheim, ed. *Welche Gesundheit wollen wir?* Suhrkamp: Frankfurt, 1995.

1 An Interview by Way of Introduction

Question: Ms. Beck-Gernsheim, you're a social scientist. At the moment you're looking into reproductive medicine and genetic engineering. What prompts you to enter the territory of the natural sciences? Isn't there an old saying, "Cobbler, stick to your last"?

Answer: That's just what I'm doing. After all I'm not discussing how high the reliability of certain procedures in genetic engineering is, or what the causes of certain errors might be, or how their reliability can be improved. These are problems for the natural scientists to solve. Instead I'm asking about the *social consequences* of such tests. For instance what happens to the people who discover, through a genetic test, that they are the carriers of a severe hereditary disease? How do they deal with this information? How will they be able to cope with it, and what kind of inroads does this piece of information make in their self-images, their plans for the future, their relationships with their partners? These are questions to which medical data offers no answers. Here we can make progress only if we begin with the social scientist's research into health—for example, with sociological investigations of the role played by affinity groups and self-help groups in dealing with illness or with psychological theories dealing with life crises, the dynamics of crises, and coping with crises.

To take another example, I'm not asking whether this or that form of hormone stimulation offers a greater chance of success for in vitro insemination. I'm asking what it means for the women involved if in vitro insemination is combined with hormone stimulation, which requires continual medical monitoring, that is, daily drives to the clinic and waiting periods there. How can women combine this with the demands of their daily professional lives, their relationships with their partners, and their other life interests? This is another question to which we find no answers in the medical literature. And this is precisely why we have to ask other questions that go beyond the medical ones—questions about the effects and side effects in social terms.

1

Question: Still, wouldn't it be better if the people who deal with such questions were able to say they had studied the natural sciences too?

Answer: Of course that would be the ideal situation. Only in that case we'd have another problem. If I first had to complete my studies in medicine and molecular biology, which would take me a number of years, when would I ever get around to posing questions about the social consequences of bioengineering? Who ever gets to that point? As we know these technologies are being developed and perfected at breathtaking speed, in many highly specialized laboratories and research institutions, not only in Germany but throughout the world. And as we also know the result is that these developments and the reports of their successes are always hot on each other's heels. Nobody can keep up with it anymore, not even the natural scientists. What gynecologist is still able to say which genetic test procedures exist today for which illnesses, and how meaningful each of these tests is? Or what human geneticist knows about GIFT, ZIFT, and the other variants of in vitro insemination? And as far as the social consequences of the new technologies are concerned, I probably would learn little or nothing about them even if I spent years studying everything the natural sciences have to offer.

Nonetheless, of course you're right—studying medicine or molecular biology would certainly be helpful to someone entering the discussion about the new biotechnologies. I would doubtless be grateful if my knowledge of the natural-scientific relationships were broader, for instance, if I knew more about the prerequisites for certain procedures. But what does it look like conversely, when the pioneers of reproductive and genetic engineering take public positions on the social consequences of the developments they are setting into motion? Some of these public statements simply make me heave a sigh. What naïveté, what narrow-mindedness! They reveal not only ignorance, barefaced and shameless, but sometimes also a downright provocative defensiveness toward anything related to the social sciences. Let me throw the question back at you: Why do people ask the social scientists to study the natural sciences first, please, before they say anything about the social dynamics of bioengineering? Where's the public outcry when the natural scientists make statements on this subject without having a glimmer of specialized knowledge, or even general knowledge, about the social sciences?

Question: That's putting it very bluntly. Is that what your experiences with natural scientists have been like? Have you encountered only lack of interest, ignorance, and incompetence when the question of social consequences arose?

Answer: No, that would be a total misconception. After all, there's no such thing as the natural scientists. Instead there are different individuals and different groups—older and younger ones—those who work in their laboratories on basic research, on very abstract and highly specialized questions, and, on the other hand, those who face the patients, with their urgent questions and wishes, in their offices day in and day out; those who strictly hold to the model of the natural sciences; and there are those who emphasize the psychosomatic interconnections. And as varied as these groups are, just as varied is their respective interest in, and readiness to think about, the social consequences of the new biotechnologies.

So please don't draw a line of battle between the natural scientists and the social scientists. The world is really not that simple. On the question of the social consequences of the new technologies, the natural scientists are by no means unanimous. Rather the assessment of these issues is a matter of very great controversy. Some scientists don't pay any attention to them at all, whereas others are looking at them very intensely and very critically.

I'd like to add that much of what I know in this area I owe to natural scientists. I've met natural scientists who were much more committed, who showed a much greater readiness to cooperate, than my colleagues in the social sciences.

Question: In this book you wish to discuss reproductive medicine and genetic engineering. Now, as every well-informed layperson knows, these are two very different realms. Why are you throwing both of them into the same pot?

Answer: In principle, these are indeed initially two separate realms. Researchers in these two areas are very quick to point this out, of course. Researchers in genetics get defensive when they are spoken of in the same breath as the "in vitro inseminators," who are doing such shady things in the area of reproduction, whereas the practitioners of reproductive medicine point out that they're only dealing with the simple desire for children, and vehemently reject anything that could lead to eugenics or the breeding of human beings.

But at the same time, any really well-informed layperson also knows that research in each of these two areas influences the other. Neither of them would have been able to achieve its great successes of recent years without the great breakthroughs that were being made in the other area. In practice there are many points of connection and interfaces between these two fields, and it's precisely their interconnection that makes their social import so explosive. To cite just one example, those who are involuntarily childless are nowadays being dragged headlong into procedures whose aim is not to have just any child but a child that is as "free of defects" as possible, genetically tested and genetically selected.

Question: The attempt to heal infertility, cure illnesses, and soothe pain is as old as the history of medicine. So why is this heated debate taking place today? Why is so much attention suddenly being paid to it in politics and the public media? Why are so many critical questions being asked today?

Answer: In the past it took a very long time to develop new methods of treatment and healing procedures to the point where they could be broadly applied. But today this process has been speeded up as though by time-lapse photography. Louise Brown, the first test-tube baby in the world, was born in 1978. A decade later there were already several thousand children who had been conceived in this manner. And I could give you many similar examples. As soon as a new breakthrough is made by a research team in the fields of reproductive medicine or genetics, it appears the next morning in headlines all over the world, and the day after that this procedure has been adopted everywhere, from Paris to Tokyo to Melbourne. This has created a totally new situation, if only because of its breathtaking speed. And that is precisely why many people feel threatened, because this speed hardly gives us time to ask questions about the side effects. For there's scarcely any time to think about whether we even want to go along with these steps and whether technological progress always means social progress.

These are not just any interventions; they're interventions that touch upon our fundamental convictions and basic values, the core of our concept of the human being and our image of the world. Obviously it's one thing to put a splint on a broken leg and another to determine deliberately the gender of our offspring or sort out embryos because they show genetic anomalies. Do we want life to become subject to our decisions, for the sake of health or perhaps only for the promise of health? Do we want life—for the sake of our desire for children, planned children, wish fulfillment—to be increasingly opened up to technology, industrialization, commercialization? These are the questions that technological progress inexorably confronts us with, which are being discussed not only in the political and public spheres but also among natural scientists, physicians, researchers, gynecologists, human geneticists, and biologists.

Question: One can't help hearing the critical tone of such questions. Does this mean that you view these new technologies primarily as a curse? Should we ban all of them as soon as possible?

Answer: I wish the answer were that simple. No, these technologies are neither a curse nor a blessing, they're both. If they were only a blessing, we could lean back, applaud and approve of them. If they were merely a curse, we'd have to put a drastic stop to them by banning them all. But because they're both at once—in another connection, Heiner Keupp coined the phrase "risky opportunities," which also fits this explosive mixture—we have to take upon ourselves the laborious task of scrutinizing them and discussing thoroughly all of their possibilities, all of their visible and hidden aspects. That is to say: What are the opportunities? What are the risks? What is permissible? What would be going too far? And last but not least, *who* will benefit from the opportunities, and *who* has to take the risks? That is, what does the social distribution of opportunities and risks look like from the viewpoint of different groups, such as men and women, parents and children, healthy people and handicapped ones, or members of the middle class and the so-called marginal groups? Who are the winners and who are the losers in this calculation of opportunities and risks?

If you're hearing a critical tone in my questions, this is bound up with the fact that many of the pioneers in the new biotechnologies are not exactly encouraging such scrupulous comparisons, to put it mildly. What they're trying to sell us, in many tones and countless variations, are always promises: helping the childless, averting illness, avoiding pain. This is a very one-sided image, with no consideration of the other side of the coin. Therefore it's necessary for this book to offer people a supplement that will round out the picture, that will deliberately take a look at the so-called "side effects" as well—all the aspects that don't fit into the list of promises. Only then will it be possible to deal openly with the issues, exchanging the various arguments and weighing them, evaluating the standpoints pro and con. Only then will democracy be possible, through public discussion of what we—this society with these basic values—consider desirable or undesirable and where we have to draw the limits.

Question: Just now you said that the calculation of the opportunities and risks would yield different results for different groups. Is this the reason why reproductive and genetic engineering are being discussed so heatedly by women in particular? What is the position of feminists— for or against?

Answer: Within the women's movement and in women's studies, the discussion is often—not always, but I'll get to that later—conducted in terms of What's in it for women? What becomes of our right to self-determination in the context of these technologies? This is the shared standard of measurement, but the answers to these questions turn out to be very contradictory. On the one side are the women who hope the new technologies will expand their options and their scope of action. For example, if a woman puts some of her ova into frozen storage early enough, she no longer needs to worry about the ticking of the biological clock, or, prenatal diagnosis relieves us of the fear of bearing a handicapped child and of the consequences that would then burden us for a lifetime. This is the position of the women who welcome—some of them directly, others despite certain reservations and restrictions—what is being offered by the new technologies. On the other side are the groups that see new dependencies, pressures, and controls being imposed on women as a result of these technologies, from the patriarchy's cravings for supremacy to the uterus as a field for experimentation where researchers can further their careers, as well as to a new exploitation of women as a class. Consequently these women advocate a radical rejection of the new technologies.

Question: Your own background is, of course, in women's studies. What is your personal position in this controversy?

Answer: To phrase it in a deliberately melodramatic way, some studies I see women as carefree consumers in the supermarket of reproductive and genetic engineering while others see them as the eternal victims of the patriarchy. Neither one of these images reflects my position. In my view the controversy is being waged on the basis of the wrong assumptions, from an overly narrow perspective. Of course it's important to ask what reproductive and genetic engineering mean for women, because it's women who have to bear the immediate consequences, whether directly through their own bodies or in their life planning and the hopes, expectations, and duties that revolve around the desire for children. There is no doubt about any of this. Yet it's still wrong, in my opinion at any rate, for people in women's studies to concern themselves only or primarily with these questions. The fact is that other groups are also affected, not just superficially but at the very core of their existence. For example, what happens to our children's right to self-determination when we take the initiative—without being able to consult the child, by definition—and make choices about the child's sex, the color of its eyes, intelligence, or whatever, whether we do this with the help of sperm donors, surrogate mothers, or prenatal diagnosis? All of these choices reflect *our* wishes, and all of them affect the bodies and the lives of these children. Finally what is the future outlook for handicapped people of both sexes? What will their life opportunities, in the literal sense, look like after we have gotten used to making genetic "defects" a standard of judgment?

Question: One could say that women's studies can't deal with everything at once, so it should probably look out for women first of all.

Answer: That depends on what you mean by women's studies. To my mind the women's movement and women's studies have not come into being only in order to struggle against the oppression of women and discrimination against them. Rather their tradition has also included developing out of this experience a sensitivity to, and solidarity with, other groups that have been oppressed, excluded, and deprived of their rights. In the discussion of reproductive and genetic engineering, this sensitivity has at times been lacking—which is perhaps understandable when we think of the speed of developments and the explosiveness, complexity, and multilayered nature of the issues that have confronted us so unexpectedly. But it's becoming evident that some groups within the women's movement and women's studies are once again thinking in terms of this tradition of sensitivity and solidarity. I'm pleased by this, and I hope very much that their voices will be heard.

Question: That would be a wish, a hope. Could you perhaps also name one important worry that comes to your mind when you look at the new biotechnologies?

Answer: If I had to pick one out, it would be a tendency that is rarely mentioned in this connection but can nonetheless be spotted everywhere if one follows the discussions. My worry is that through the biotechnologies we are becoming increasingly dependent on the experts, or even on an "expertocracy"—that is, a society in which ultimately only the experts rule. Please don't misunderstand me. This trend toward expertocracy is certainly not a development that has been solely created by the biotechnologies. There are many other causes that are inherent in the project of modernism. But the biotechnologies are helping to intensify this trend; they're giving it yet another strong impulse. We're already experiencing in this field the endless chain of pro and con arguments between experts and counterexperts. The questions that have to be answered are obviously extremely complicated and include many dimensions; moreover they require a knowledge of the natural sciences. Who has the competence nowadays to deal with them? In many public discussions I've seen how overtaxed and helpless the listeners felt in the face of the discourse of the experts that was rolling down from the podium and washing over them—in the face of the intimidating procession of technical expressions, foreign words, and specialists' jargon. In some private conversations I've seen politicians or university colleagues admit defeat regarding these questions; they were unable to find any pathway through the jungle of contradictory information. And how must those people feel who never had the opportunity to attend college and perhaps only went to high school? Or what about the foreigners in our country, especially those who come from different cultures, have little schooling, and speak only a broken version of our language?

I said that we are experiencing the trend toward an expertocracy in many other areas as well. But here this trend causes especial concern, because the issues in this field directly affect our image of the world, our image of human beings, our basic values—affect them not through many intermediate steps but directly in our own lives, in decisions that some people may be facing sooner than they expect. For example I look at the women students who sit in front of me in my seminars. They're in their early twenties or midtwenties or sometimes older. Most of them will go to college for a few years, then try to establish careers, and at some point many of them will want to have children. Some of them will then be forced to realize that it doesn't always happen immediately the way they had hoped. Many others will become mothers late in life. And thus many of them—one knows this in advance—will be confronted by the question of which offerings of the new technologies they want to take advantage and which ones they would rather not get involved with. These are not abstract, theoretical topics. They're decisions that will make a direct impact on their bodies, their lives, their partnerships, and their plans for the future.

So this is where the problem lies, and I have no patent remedy for it. Is there a path that will lead us out of the dictatorship of the experts and counterexperts, out of the expertocracy where the doctors, geneticists, biologists, sociologists, psychologists, jurists, and specialists in ethics lay out their pros and cons before us in many loops of argumentation, while all of the others, the people in the outside world, trust that the experts will know where they're leading us or simply follow the course that seems plausible at a given time or happens to be in the headlines? In other words how can we keep democracy alive in this area—in decisions that are so fundamental and that determine the course of the future?

This interview with Elisabeth Beck-Gernsheim was conducted by Elisabeth Beck-Gernsheim. The guiding thread of the interview was the method of realistic constructivism.[1]

2 The Technologies as Seen by the Social Sciences

Very rapid developments have taken place in reproductive medicine, biology, and genetics in recent years. These include the perfecting of artificial insemination by means of deep-freeze techniques and sperm banks, insemination in the test-tube and embryo transfer, and, finally, the new possibilities of genetic analysis, from prenatal diagnosis to predictive medicine and preimplantation diagnostics.

These technologies jointly are making possible entirely new forms of intervention in the area of reproduction—in the very substance of human life. This is why they have become the object of numerous discussions among scientists, politicians, and the general public. Both advocates and critics see that we are poised to make a "qualitative leap": the very creation of human beings is becoming feasible. This means that reproductive and genetic engineering are increasingly taking on the role of creators, and entirely new questions are presenting themselves. What criteria should be used to make decisions that affect the creation of new life? And who should be making these decisions? How, and to what end, should we use this power, which is growing for us and beyond us?[1]

From the perspective of research in the natural sciences, these developments are evaluated according to their biological or medical results and the corresponding calculation of risks versus benefits. For example, How high is the success rate of in vitro insemination? How frequently does this result in multiple births? How great is the likelihood that prenatal diagnosis will discover an anomaly such as Down's syndrome?

These are the intrinsically medical questions that are part of the conventional paradigm of research in the natural sciences. But the dramatically new aspect of reproductive and genetic engineering—and this is the essence of their fascination, their promise, and their menace—is

11

that they are fraught with consequences which are causing far-reaching changes in society, social institutions, politics, and the individual psyche.

Under these circumstances a reorientation is becoming necessary. Precisely because the new technologies have such enormous potential and because their effects are so much more far-reaching than the conventional procedures of medical diagnostics and therapy—for just these reasons it is an *anachronism* to look only at their medical results and ignore all of their broader consequences and side effects. This abbreviated and one-sided perspective reflects what the American social scientist Neil Postman once called the "triumph of one-eyed technology." He calls technology one-eyed "because it, like the Cyclops, sees only what is standing directly in front of it." And this, according to Postman, is inherent in the veritable basic principle of technology: "Who would seriously expect a machine to think about its own side effects, or to worry about the social and psychic consequences of its own existence? Machines don't ask questions, they don't notice their surroundings. They perceive the future from the rigid perspective of their technical possibilities."[2]

1. The Provocation

The idea that Postman vividly formulates points directly at the core of the many controversies we are experiencing today in politics, the media, and public life. It is symptomatic of a new round in society's perception of technology and its consequences. Instead of a debate concerning the empirical biological risks, we are increasingly seeing a comprehensive reassessment of technology that takes account of risks in the areas of social policy and culture, illuminates the goals and concepts of research, and refuses to relinquish evaluation of the political aspects entirely to the researchers.

And here lies the stumbling block. For many natural scientists, questions about the broader implications of their research and behavior—the societal and social-policy consequences, the political and economic interconnections—are still a provocation: alienating, irritating, and threatening. To many such questions seem an inadmissible intrusion into their own territory. This is wholly understandable in terms of their experience. After all the natural sciences and technology were both supported for a long time by a faith in progress that seemed self-evident and in which their legitimation was grounded. This faith in progress, which was shared by most groups within our society, was based on the assumption that the results of the activities of the natural sciences and technology, that is, the growing control over nature, promoted the common good as well as individual well-being. At that time most people believed, usually without reservation, that we would raise productivity and increase prosperity, protect people from illness, and prevent starvation. This belief was typified by the assumption of harmony: in this view, progress in the natural sciences was synonymous with progress toward the general good.

And are we now supposed to believe that all this is no longer valid? Controversies have arisen in politics and public life, within the sciences and among them, and within the natural sciences themselves, from the discussion of the destruction of the natural environment to the discussion of the increasingly extensive incursions into human nature. What is becoming visible is the end of the "consensus on progress" that has long been the foundation of industrialized society.

This collapse of previous lines of argument confronts the researchers with a wholly new situation. For many of them, especially those of the older generation, this removes the formerly self-evident basis of their daily activities and calls into question their life's work itself. They are being faced with a reevaluation of their public image. Once regarded as saviors of mankind, they now are often viewed as destroyers of nature and the environment, blindly endangering the future of our planet. They are being faced with questions that go far beyond the categories of their training and professional specialization—questions that affect their professional self-image and professional identity. This is bound to cause insecurity and in some cases bitterness leading to hostility. What makes these confrontations even harsher—including those within the natural sciences, especially within the younger generation—is the presence of increasingly large groups that reproach the traditional natural sciences with reckless endangerment and demand a "different" conception of natural science.

What is now moving from the various sidelines onto center stage is the question of the *social acceptability* of the results produced by the activities of the natural sciences and technology. Thus the perspective changes: people are now deliberately asking about the so-called *side effects* of progress, the risks and potential dangers it implies. People are calling for a type of technical research that recognizes such potential dangers ahead of time and makes them accessible to public discussion.

This applies to many areas, from nuclear energy to bioengineering. Let's take the example of genetic engineering. "In the critical discussion of the unexplored potential dangers of genetic engineering, the issue is not only technical acceptance in the narrow sense but also the question of the ethical and social acceptability of this type of bioengineering."[3] It is becoming obvious that genetic engineering is not only providing us with the technical means of attaining technically defined goals, but it is also, in a much more far-reaching process, initiating a new relationship of human beings to themselves and their own nature. This process necessarily raises fundamental issues that require renewed reflection on the meaning and purpose of such interventions.

> The decision for or against the use of genetic engineering actually has strategic significance not on the level of individual areas of policy . . . but on the level of culture as a whole. Life, which until now has been surrounded by at least the remnants of an almost religious untouchability, is becoming as technically manipulable as plastic. Is this appropriate? Is a system of ethics, are individual values, which permit or virtually demand such a technologization, appropriate? . . . Can we, and may we, "emancipate" ourselves at will from the bases of biological evolution? . . . These questions . . . touch upon the meaning of our existence and our behavior; they affect the basic assumptions underlying our ways of thinking, our relationship to nature, and our moral institutions.[4]

Given the enormous possibilities for intervention that are being opened up by bioengineering, the world is becoming a laboratory and society itself is becoming the object of the experiment. The inevitable questions that arise are: *What kind of future do we want? In what kind of society do we want to live?*

2. Prerequisites and Presuppositions—A Practical Selection

When the discussion turns to topics such as these, prerequisites and presuppositions and premises and priorities must be redefined. Those who are used to thinking and acting in terms of the paradigm of progress find themselves suddenly on unexplored terrain, bereft of their old certainties. The tension intensifies; some grow more impatient, others more rigid.

In order to get beyond the positions in which we are bogged down and in order to make a dialogue possible, it is first of all necessary to identify clearly the differences. This is precisely what the following key words will try to do—they are meant to convey an idea of how the new view of technology is changing our field of vision. Perhaps this will clear up a few of the misunderstandings that keep cropping up in the dialogue between the natural scientists and the social scientists, creating endless irritation. Perhaps—this, too, could happen—the dispute will then become even more bitter. But then at least we will know what we are arguing about.

Intentions Are Different from Consequences

In the new discussion the natural scientists, who have been shaped by the paradigm of progress, often feel misjudged. They point to the intentions that motivate their research. Finding a new medicine, a better vaccine, more effective possibilities for prevention and therapy—Aren't these goals important and right? So why doesn't anybody have faith in their good intentions?

There is a misunderstanding here, a major one. When social scientists assess consequences, their aim is certainly not to impute bad intentions to all researchers. They are only, literally, evaluating *consequences*. And this requires them to differentiate between the intentions and the consequences of an action. The result that was aimed at isn't always achieved, and sometimes even a very different result comes out in the end. This applies to all areas of human activity, from politics to love to child-rearing (many parents would love to raise a little Einstein, but only a few succeed in doing so). Norbert Elias once put this quite succinctly: "Under no circumstances should one represent the development of society as though everything proceeded from the desires and plans of human beings. Out of the intentions of many individuals, which in many cases cancel each other out, arises something that may be entirely different from what they want."[5] This sentence applies not only to the development of society but also to the development of science, research, and technology. A comprehensive study of the use of the knowledge gained from the social sciences comes to the provocative conclusion: "The use of the results has nothing to do with the results that are used."[6] We can safely assume that the situation is not wholly different in the natural sciences.

In concrete terms this means that the motives of the scientists and researchers and the doctors and counselors who work in the field of, say, human genetics, may be entirely respectable, honorable, and committed to the values of the medical profession and the Constitution. Nonetheless, when the social scientists assess consequences, they are not interested in the intentions of individuals but in "the results that come out" in the social sphere, that is, how people's behavior changes when new possibilities of genetic diagnostics become available and collide with the limiting conditions of society and how in the course of this interplay new ways of structuring individual lives then arise and become fixed. So the question here concerns the actual use of human genetics within the framework of social, political, and economic conditions, in interaction with the influence of diverse interests, competitive strategies, power constellations, and conflicts between a great variety of groups and institutions (from policy makers and parties to churches, labor unions, and other associations).

Normative Goals and Complex Motives

In the natural sciences it is always the goals of in vitro insemination or prenatal diagnosis or genome analysis that are referred to. If we look more closely at this way of speaking, these goals result from the tasks set by the medical profession in terms of the norms of our legal system and our society—in concrete terms, the early detection of illness, prevention, and therapy. But this way of speaking is admittedly an abbreviated one in that it remains normative and thus ignores an essential condition of social reality: it ignores the diversity of the groups that deal with the development, application, and acceptance of, for example, procedures for genetic diagnostics—researchers, doctors, patients, businesspeople, etc.—and the fact that these groups often have very different motives, interests, hopes, and expectations regarding genome analysis. In many cases these individual motives may be compatible with the official, socially legitimate goals of genome analysis. But at other points incompatible interests or even clear conflicts of interest may arise. For this reason it would be a grave mistake to draw direct inferences from the normative goals to the way these procedures are actually used. Let me give just a few examples.

People who work professionally in the field of reproductive and genetic engineering are no different from other professional groups: they bring into their professional situation individual interests oriented toward securing their economic and social status.[7] This means that researchers and physicians, like other professionals, do not do their work solely out of charitable motives, to help mankind and prevent suffering. They also want to have an income, secure their social positions, possibly improve themselves through promotions, and perhaps even have brilliant careers and win the Nobel Prize. Obviously such motives are not the specialty of people who work in reproductive and genetic engineering; they can be found in all professional groups (yes, even among social scientists), and they are by no means illegitimate (who would think of demanding that all doctors be forced to apply for social welfare?).

But the essential point here is that such motives (not just those set as norms) also help to determine the development and application of reproductive and genetic engineering. Thus, for example, the indication for genetic-diagnostic analyses during pregnancy may be determined according to more generous criteria in situations where the doctor in question has access to the necessary laboratory installations and profits from their use. And the more doctors go into the field of genetic diagnostics—because this is an expanding field that still offers opportunities for earning well despite the general surplus of doctors—the more likely it is that their readiness to diagnose such indications will increase. (It is already becoming evident that the age limit for amnio-centesis is sinking in tandem with the increase in laboratory capacity.)

Moreover it is important for the social utilization of human genetics that the researchers and physicians working in this field are not a homogeneous group but are very heterogeneous with regard to their motives, orientations, and personal values. How greatly their positions may contrast can be seen especially in the controversies kindled by the catchword *eugenics*. There are many scientists, especially in Germany, who explicitly and unequivocally reject any kind of eugenic orientation. And there are many scientists, especially in other countries, who just as explicitly and clearly defend eugenic goals. This means, to begin with, that human genetics "as such" can be used to serve different goals. And it means, furthermore, that it would be false to conclude from the basic orientation of the opponents of eugenics that the future use of human genetics—nationally and even more so at the international level—will take place only within the boundaries they have set and recognized. Things that for some scientists approach taboos might be viewed by others as still permissible, desirable, or even morally necessary—and the latter group of scientists will act accordingly.

Finally the patients and clients themselves also bring their own interests into medical technology. For instance it is surely a woman's hope of having a child that brings her into a doctor's office to discuss this wish. But what is formulated as the desire for a child often is related to entirely different motives, which are sometimes even the primary ones and which lie beyond the scope of the doctor's competence—for example the hope of mending a marriage or consolidating a shaky sense of self-esteem.[8] Whereas in a normal situation prenatal diagnosis is sought in order to dispel anxiety about a possible handicap, here it might be sought in the hope that a genetic defect *will* be discovered, so that the woman may receive an indication for an abortion by legal means.[9] This is of course an extreme case. But it is normal, and not an extreme case, for human beings to have many layered motives which—to put it mildly—do not always coincide with the preachments of official norms and models. ("Man is a crooked piece of timber," as Kant already knew.) Thus constellations may arise where the patient's wishes and expectations by no means lie within the normative assumptions and limits inherent in the medical profession's self-image. To mention some experiences from the field of prenatal diagnostics:

> frequently enlightened, well-informed parents make demands, backing them up with arguments, that go far beyond what the doctor can accept responsibility for and what was originally intended. Because in our pluralistic society we are used to accepting the most varied levels of justification for people's behavior, we should not be surprised when parents decide to test the unborn child for entirely different reasons than those which their counselors would like to concede to them.[10]

Since Freud we have been able to read up on the fact that the human soul is a labyrinth with many winding paths where longing and pain, desire and suffering merge in many a strange combination, driving people into all kinds of complications. But in the world that is presumed by the pioneers of bioengineering, none of these complications exist. Here there are only defects, anomalies, findings—but not the human beings they pertain to. There is no soul, and even more certainly no complications, no secrets, no variety, no coexistence of varied motives, no reasons behind the reasons. Only the objective data are talked about (e.g. the success rate of in vitro insemination or the likelihood of a genetic defect when risk carriers are involved), but no attention at all is paid to the ways in which these act upon subjective expectations, hopes, and fears (e.g., on the researchers' ambitions or the desperate wishes of the patients) and how people's actions arise out of this connection. All of this remains outside the discussion, and it is explained away as an "untypical situation" or called "irrational," which means we don't want to see it. The human being as envisioned by the pioneers of bioengineering is smooth-surfaced, free of feelings, sterile—a laboratory product.

Is Technology Destiny?

A widespread attitude is that technology to a certain extent is destiny. In this view technical progress and the emergence of new technologies have an inexorable power. Developments wash over us in the same way natural forces do. We cannot evade them or change them in any way; we can only adapt to them.

But this view is incorrect, for the acceptance of new technologies can never be understood from a solely technological perspective. Whether, how, and by whom a new technology is used depends on many circumstances, especially on culturally and historically reinforced values, socially accepted norms and models, the parameters set by the legal system, the images projected by the mass media and advertising, the discourse of experts in public forums, and, finally, the anticipated financial costs and benefits. Let us take prenatal diagnostics once again as an example (it will be preimplantation diagnostics in the future). Among the important conditions influencing the demand for it and its use are:

- the current leading values of "health" and "consciously responsible parenthood";

- the change in family structures, namely, the nuclear family of today, the drop in the birth rate, the increasing number of working women (under these circumstances it is becoming more difficult to accept a handicapped child and thus more thoughts and wishes revolve around the child's health);

- the institutional services on offer for the care and education of handicapped children—the fewer services are offered and the less adequately they are equipped, the more families feel overwhelmed at the thought of having a handicapped child;

- images in the mass media, but also sometimes in science, which arouse false expectations as though illness were avoidable with the help of the new procedures and as though a child's health could be guaranteed;

- the discussion of the cost explosion in health care, that is, the prevention of genetically influenced illnesses thus becomes a legitimate goal of social policy;

- the parameters set by the legal system (e.g., the Constitution, recognition of the right to self-determination, etc.). Here the most important aspect is the regulation of abortion, that is, whether prenatal diagnosis is associated with "opportunities for action" if the results are negative;

- the life history of the woman/the couple: Was the pregnancy intensely longed for or was it rejected? Is there a religious orientation, and, if so, how strong is it? Have the two potential parents ever lived with handicapped people, or do they know any handicapped people through their surroundings?

Thus biotechnology is certainly not "destiny": it does not dictate whether or how it is to be used. But on the other hand, neither does it remain neutral in the context of society. It offers new possibilities for dealing with health, illness, and handicaps, which interact with cultural values, accepted guidelines, legal regulations, etc., and thus it competes with previously accepted patterns of behavior and may even help to displace them. In this sense technology can be understood as a spiral-shaped process. "Technology appears as the product and the instrument of social needs, interests, and conflicts; technology is simultaneously a cause and an effect."[11] It arises from a certain socio-cultural background, and in the process of being applied it changes this background.

In the following chapters I will try to illustrate this spiral-shaped process. To use Postman's phrase, I will try to make it two-eyed rather than one-eyed and to demonstrate the *social dynamics set into motion by reproductive and genetic engineering*, that is, the fundamental change in expectations and standards of judgment, wishes, and pressures in the highly sensitive area that includes reproduction and the desire for children, health, illness, and handicaps.

3 From the Pill to the Test-Tube Baby: New Options, New Pressures in Reproductive Behavior

This chapter[1] focuses on the new techniques of reproductive engineering and their social and psychic consequences for women and men. Here I will use the concept of "reproductive engineering" in a very extended sense: it will encompass all the forms of biomedical intervention and assistance that are available today for ensuring—or preventing—the birth of a child. Thus the spectrum ranges from the new forms of birth control, which offer nearly 100 percent certain contraception (the "pill"), to the spectacular new forms of treatment for infertility, such as artificial insemination ("sperm bank," "deep-frozen sperm") and in vitro insemination.

I will approach this subject in two steps. First the arguments of the pioneers of reproductive engineering will be compared with those of their critics. Here the context of the application and acceptance of technologies comes into play, together with the momentum they develop and the side effects they cause. This creates the frame of reference for a discussion of two concrete examples from the repertoire of reproductive engineering: oral contraceptives and the new methods of treating infertility. In each case the question is What new possibilities for action and what new pressures to act are associated with them? How are these technologies affecting the lives of women and men, the ways they live together, their wishes and plans, their hopes and fears?

1. On Momentum and Side Effects

What the new techniques of reproductive engineering have in common is the fact that they are increasingly detaching sexuality and reproduction from the pressures and limitations of nature. A central area of human life, which in the past could be influenced only to a very limited extent, is today becoming increasingly subject to our plans and actions and dependent on our decisions.

The question thus arises of how to assess the consequences of this development. The pioneers of the new reproductive engineering emphasize the progress associated with it, as well as the extension of our freedoms, options, and opportunities to make decisions. Their arguments are as reasonable as they are plausible: women and men who do not want children need not forgo sexuality or live with the risk of an unwanted pregnancy. Women and men who have remained childless without wanting to are still able, with medical assistance, to have children.

On the other hand in recent years a broad coalition of scientists from a great variety of disciplines—the spectrum stretches from philosophy to medicine, from theology to women's studies—has arisen that regards the new developments from a far more critical standpoint. Above all, advocates of this view point to the unplanned and unwanted "side effects" of technological intervention. Their arguments run as follows: the development of the new reproductive technologies not only produces new options, it also eradicates old ones; it creates not only freedoms, but also new pressures, controls, and dependencies. Moreover technologies are never neutral with regard to society: they have a retroactive effect on the situation within which a decision is made, on alternatives for action and standards of judgment; they change individual expectations and patterns of behavior, social norms, and standards. To modify a phrase of Max Weber's, technological development is not a "cab you can get out of whenever you please."

Now the advocates of the new reproductive technologies are by no means denying that they may also entail risks.[2] But they reply that if a person judges the risks to be too high, he/she is certainly not obligated to take advantage of these medical opportunities. In other words nobody is being forced to use them.[3] They also admit that here, as with other developments, use and abuse lie close together. But they point at the individual's personal responsibility and decisions as a way out of this dilemma. It is compressed into a formula: "Nobody is forcing us to abuse them."[4]

The critics of the new reproductive engineering, in turn, object that it is one-sided and an oversimplification to regard responsibility and freedom of decision as solely individual attributes. Individual decisions, they argue, are always part of a social system that rewards certain possible choices and punishes others. Moreover choices are always overdetermined by social processes: the processes of making something public knowledge, transmitting information, perceiving and defining problems, and establishing norms, directly and indirectly. So our attention shifts to the social process of applying technologies and gaining acceptance of them, with its inherent momentum and impact.

Thus at the center of the controversy is the question of how we should understand the march of progress: as individual freedom of decision or as a socially determined constraint. The following section will sketch out in broad strokes some of the arguments being raised in this connection.

Gradual Acceptance, or, A Quiet Revolution

In many cases the acceptance of new technologies takes place in small, and thus barely noticeable, individual steps which nonetheless systematically build upon each other and thus continually move the development along. In such a process, no far-reaching change or qualitative leap is visible at any given point—yet a fundamental change has taken place in the end. What is happening before our very eyes, although we don't see it, is a "quiet revolution."

> The genetic technologies, which are so revolutionary from the scientific point of view, (can be) joined up, in their practical application, to already established procedures without any problems and almost imperceptibly. They are introduced in small steps, each of which leads beyond what went before, but is covered by analogy with what is already known. Each of these steps, viewed in itself, is plausible.[5]

> The process . . . continues along a hundred pathways and through a thousand small steps, everywhere full of unknowns with regard to the critical threshold values—that is, open questions as to how far one may go here or there; not through dramatic decisions but through banal normality, and through inherently innocent means . . . which are beneficial to life.[6]

In the sixties and early seventies, for example, the use of artificial insemination—a technically uncomplicated procedure—increased. This helped to break down emotional barriers regarding intervention in the area of reproduction and created precedents for juridically dealing with children having "stranger" genes. Thus in a very concrete way artificial insemination also smoothed the path for the eventual use, as substitutes, of the ova and/or uteruses of other women. Further relatively uncomplicated technical procedures, which were separately developed, are the hormonal stimulation of the ovaries and laparoscopic surgery, both of which legitimated interventions in women's reproductive organs that were not dictated by illness. All of this has resulted not from a conspiracy but from a continuous process of gradual and partial changes, which, because of its particular nature, generally remains invisible. Every change in medical practice has led to an alteration of social relationships and expectations, but none of these changes has been dramatic enough, taken in itself, to spark controversies in its initial phase.

Uncontrolled Application, or, Medicine as Subpolitics

Intrinsic to the self-image of democratically structured societies is the fact that central questions affecting the future of this society are dealt with publicly in a process of sounding out the people's political will. In actuality, however, the development of modern technology has initiated processes that result in central questions about the structuring of human coexistence, which are being more or less settled via the back door of technical parameters, a long time before parliaments, parties, or other democratic institutions can exert any influence upon them. Technology, medicine, and, above all, the new reproductive technologies are becoming the vehicles of an uncontrolled "subpolitics."

Through the structures of action inherent in it, medicine has access to . . . a carte blanche for implementing and testing its "innovations." It is always able, by means of a policy of "faits accomplis," to undermine public criticism and debates on what a researcher may or may not do. . . . In terms of social structure, in the sub-politics of medicine there is no parliament or executive branch in which the decision could be investigated ahead of time with regard to its possible consequences. There is not even a social institution for making decisions. . . . We must bear this in mind: in the thoroughly bureaucratized, developed democracies of the West, everything is scrutinized to determine its legality, jurisdiction, and democratic legitimacy, while at the same time it is possible in everyday life, outside parliamentary institutions, to annul the foundations of life as it has been known and lived hitherto, evading all bureaucratic and democratic control mechanisms in a closed decision-making process despite the growing torrent of general criticism and skepticism."[7]

Reproductive engineering offers many a vivid example of this development. The commissions that are supposed to regulate its application are typically appointed only after the fait accompli—for example, after the first test-tube baby has been created and so on. Thus a crucial step has already been made toward letting the situation develop its own momentum, for such commissions' mode of operation is based on an implicit model of progress in which it is not innovation but the blockage of innovation that has to be justified. Yet, in the given circumstances, it is hardly possible to find an unambiguous and generally accepted criterion for justifying the blockage of innovation.

Where would a veto of this sort come from? The jurists measure innovation according to the standard of legality. So are the possibilities opened up by reproductive engineering compatible with, for instance, the regulations laid down by the Constitution? Do surrogate mothers or test-tube reproduction violate human dignity? Such questions lead us into complicated processes of comparison. There is a wide gap between the possibilities the lawgivers had in mind and the possibilities the pioneers of reproductive engineering have granted us. Thus in practice there is a great deal of leeway for interpreting and construing, and this, in turn, results in the fact that different jurists have arrived at different assessments—as was very obvious at the German Jurists' Congress in 1986, for instance.[8] But this leeway for interpretation is far more likely to favor the acceptance of innovation than its rejection, because, as the jurist Mayer-Maly writes, "Partial openness to changes in values is inherent in the structure of the law." If there is a change in the framework set by technology and civilization, a change of conviction may also take place, and the law must then adapt to it.

The situation of theologians and ethicists is similar. Here, too, there is a wide gap between the general principles of religion and ethics and the recent achievements of natural science. For example, which passages of the Bible can be reconciled with in vitro insemination, and how? Is the latter reprehensible because it requires masturbation, or is it permitted because it serves to fulfill the ancient commandment to "be fruitful and multiply"? Or must we make a distinction between its use in homologous and in heterologous systems? There are many interpretations, controversies, and divergent positions within religions as well as between different religions; similarly there are controversial statements and standpoints in the field of ethics. But wherever so much leeway is left to uncertainty, one can expect that in the further course of things the force of the real will establish itself and innovation will be accepted. To play on Shakespeare, "what you will" is what you get.

Technological Colonialism, or, The Suppression of Other Views and Values

The offerings of the new technology are part of the comprehensive process of rationalization inherent in the project of modernism, whose aim is to "scientize" ever broader areas of behavior. In the course of this development, habitual patterns of behavior are becoming increasingly suspect. The "demystification of the world" (Max Weber) continues. Nowadays whatever is new and modern and comes to us with the conviction of science is judged to be right. Given these premises our field of action has been redefined, and this brings with it a host of consequences. "To be sure, science always opens up new possibilities for action, but it closes others. Those which science itself opens up are judged to be rational, and it becomes extremely difficult to apply to them any appeal to ethics."[9]

As we know this development is certainly not proceeding in a straight line. Different images are created depending on the social group, social class, or milieu. Patterns of life planning, that is, the controlled and disciplined conduct of individual lives, which are well established within a given group, are just becoming visible on the horizon for other groups. Accordingly the offerings of reproductive engineering will not find the same degree of acceptance everywhere. Some groups' readiness to use it is growing quickly, whereas in other groups mistrust and hesitation will predominate initially. Prenatal diagnosis provides a vivid example: empirical data show that middle-class women who have good educations and good jobs are primarily taking advantage of this opportunity. And women from the working class, immigrant groups, and ethnic minorities tend to have reservations about prenatal diagnosis.[10]

But if this information is accurate, then we must conclude that the new technologies are socially biased rather than neutral—and the fact that this is a covert bias does not make it any less effective. In other words the new technologies are becoming the allies of those who have most effectively internalized in their lifestyle the commandments of rationalization and planning—those who ambitiously build up their professional careers, effectively use preventive health care, and are well informed about savings and loan associations and pension plans. As we know these are the members of the middle class, who have professional qualifications and have attained a certain level of material well-being.

But what about the members of the so-called marginal groups who are less adapted to the lifestyle of rationalization? Presumably they will also remain behind in the process of behavioral change that is inherent in reproductive and genetic engineering, that is, the process of dealing with one's own body and life in a conscious (controlled, disciplined) way. They will thus come to look even more like unenlightened and suspect outsiders. They will probably be subjected to sanctions and gentle or less gentle pressures to force them to follow the path of progress. There is no lack of historical examples such as the efforts made in the late nineteenth century by doctors and women of the upper middle class to teach working-class women the rules of hygiene and infant care. At that time detention "for the detainee's own good" was the result.[11] Similar tendencies may soon appear in the field of reproduction, contraception, and preventive measures during pregnancy. Initial examples of this in the United States are already well known: cuts in social welfare if poor people "recklessly" bear children instead of acting "responsibly," no insurance payments for a handicapped child if the mother did not undergo prenatal diagnosis, and criminal proceedings to control the behavior of pregnant women. To be sure these measures are formulated in general terms, but in practice they very clearly apply to "backward" marginal groups. (For example, in 80 percent of the cases where a judge orders a Caesarean section against the will of a pregnant woman, the woman is Black or Asiatic.[12]) This situation is not the result of a sinister conspiracy; on the contrary it is often associated with socially minded motives—a tangle of ruling interests and benevolent intentions, support from the welfare state, and bureaucratic regulations.

The explosiveness of this situation can be seen especially clearly in the USA, which has a multitude of ethnic and cultural minorities. This fact creates many lines of conflict in the area of genetic counseling: "Gentle or not-so-gentle forms of pressure are exerted on the women who come to genetic-counseling sessions, depending on their socio-economic status. The reproductive technologies offered within the framework of state-financed programs are perceived by poor and colored women as part of the historical tradition of eugenics and population control. In many cases, the meanings and values associated with having children and with abortion are played out between a well-educated white genetic counselor and a poor woman of black or Hispanic origins, against a background of mistrust, control, and unequal distribution of power."

For women who are members of a minority group or have little money, the following conflict occurs especially often: "In many cultures, people believe they have little control over their destinies or that they must not interfere with God's will. For example, many people from Southeast Asia believe that amniocentesis interferes with the process of natural selection of the human population, which they consider holy. Many religious Catholics view handicaps as acts of divine providence. Instead of assuming that all cultures share the same attitudes, it is important to recognize what a huge variety of cultures exist."[13]

The General Nature of Acceptance, or, All of Us Are Affected

"The social problems entailed by the dynamics of technology result from the expansion of legitimate leeways for action."[14] Wherever new possibilities for action are opened up, standards of behavior begin to change. Things that in the past were first regarded as impossible and then as sinful become, in the present, first novel and then normal, and in the future they might even be dictated by law. A variety of conditions contribute to this process.

"Appetite (is) stimulated by possibility."[15] It is in the nature of the new technologies to create new needs, thus constantly expanding their field of application, their "market."

Familiar, traditionally aspired-to goals may be better satisfied by means of the new technologies whose emergence they have inspired. But the reverse may also happen—and is becoming more and more typical: new technologies may inspire or create, or even force upon us, new goals that nobody has ever thought of before, simply by offering us their feasibility. Who ever wished to have presented to him, in his living room, great operas or open-heart surgery or the recovery of corpses from a plane crash (not to mention the ads for soap, refrigerators, and sanitary napkins that go with them)? or to drink his coffee out of disposable paper cups? or to have artificial insemination, test-tube babies, and surrogate pregnancies? or to see clones of himself and others running around?

Thus technology adds new and novel objects of human desire and need—indeed, whole species of such objects—to those that already exist, and in doing so it also multiplies the tasks it is expected to do. The latter point shows the dialectic or circular aspect of this situation: goals that initially were not asked for and were perhaps accidentally generated by acts of technical invention, become necessities of life once they have become part of the habitual socio-economic diet. . . . Thus "progress" is not an ideological ornamentation of modern technology, nor is it merely an option offered by it which we can choose if we want to; it is a drive that is inherent in modern technology itself.[16]

"The long-term interest of reproductive medicine is to transform the techniques of healing into general procedures of procreation."[17] A pattern is starting to form: new biomedical aids are introduced in order to prevent or relieve suffering within a narrowly defined range of unambiguous "problem cases." Then a transitional or habituation phase sets in, during which the field of application is continually expanded. The final stage can be foreseen: all women and men will be defined as potential clients—now, of course, not to prevent direct harm to their health, but because of the "effectiveness advantage" of technical intervention over nature's accidents, unpredictability, and susceptibility to interference. Thus numerous forms of surgical intervention in women's reproductive organs have been expanded as a "market" and have increasingly become routine measures that are taken for granted. Uterectomy, Caesarean section, and episiotomy are familiar examples. And a similar marketing process is in the offing for the new techniques of reproductive engineering.

For example in the past few years the indications for in vitro insemination have multiplied and become blurred—in fact, they are now practically unlimited.[18] Initially in vitro insemination was a very special procedure for the benefit of women whose ovaries were missing or obstructed. But now in vitro insemination is also being used by couples where the woman's reproductive organs are completely healthy but the quality of the man's sperm is unsatisfactory. In vitro insemination is even being offered as a "last chance" procedure—that is, to all couples whose infertility has not been medically clarified as to its causes. And this development is continuing. Some experts are already portraying in vitro insemination coupled with embryo assessment as the ideal method of the future for attaining early diagnosis and preventing genetic anomalies.[19] Already the director of the embryo-transfer team in Kiel has declared that problems are caused by the fact that not only superior but also inferior sperm cells are used by human beings. If the quality norms of cattle production were applied, he explained, they would be fulfilled by only one man in ten (this was followed by the remark that bulls having inferior sperm quality end up in the slaughterhouse).[20] Already proposals are being made that in vitro insemination and frozen embryos be used to enable couples to plan precisely the intervals between their children.[21] Already an increasing number of men are depositing sperm cells in sperm banks before being sterilized—"just in case" they should one day change their minds.[22] Already married couples are asking whether a man other than the husband could donate sperm, because the couple is dissatisfied with the appearance or personality of the husband. Similarly women have asked whether they could use the ova of other women, because they were dissatisfied with themselves in some respect.[23]

The road is leading us from the treatment of infertility toward the rationalization of reproduction. In the process "nobody is forcing us to abuse" anything, as the advocates of these technologies assure us. This may be true (at the moment, in any case). But the crucial point lies elsewhere: the boundaries between use and abuse are blurring and changing in ways we cannot easily monitor. Who can be expected to make decisions about what is still "appropriate" and what is "going too far"?

Acceptance under Pressure, or, From Promise to Compulsion?

"The use of these technologies by no means remains merely an option for the individual. It turns against him as a vehicle of social compulsion."[24] What is becoming apparent here is the vision of a distant—or, then again, perhaps not so very distant—future in which the new possibilities are transformed into their opposites: direct or indirect compulsions.

> Once this or that possibility has been . . . opened up and developed on a small scale through practice, it tends to force upon us its application on a large and ever larger scale, and to make this application a permanent necessity of life. . . . With every new step (= "Progress") of large-scale technology we subject ourselves to the compulsion to take the next step, and bequeath the same compulsion to the world that will follow us.[25]

> The questions about the new reproductive engineering are significant for all women. . . . Whether we want children or would like to stay childless; whether we are beyond childbearing age; no matter how we live out our sexuality—all of us run the risk of becoming test-tube women. All of us run the risk of being subjected to a host of controls, beginning with technological interventions when we are pregnant, progressing through legal regulations that declare the foetus and the woman carrying it to be two separate "patients," to job regulations that force female employees to have themselves sterilized.[26]

Any reader who believes this picture is too pessimistic should consider the following examples. One of the promises being offered to us today is the vision of some researchers who extol the reproductive technologies as the royal road to family planning: "It may soon be possible for parents to carry out total family planning, from controlling the size of their families to the sex of their offspring and the sequence of male and female children."[27] Similarly there are popular science books that view bioengineering as a great opportunity for self-determination and freedom:

Modern science promises to place in our hands the definition of what a mother, a father, a family is, even of what human life is. We will decide not only whether and when, but also whose ova or sperm cells will be used, where insemination will take place, in whose belly the foetus will grow until its birth, of what sex it will be, what defects will require abortion or correction, and finally also what genetic improvements we want with regard to intelligence, character, and appearance.[28]

Admittedly such promises leave an unpleasant aftertaste. For the history of technology has also taught us how quickly possibilities turn into their opposites—compulsions. Thus total family planning could become the supervised and socially regulated family. Here there is a whole range of gentle and not so gentle interventions: from the "myth of voluntariness"[29]—that is, data manipulation, biased "information," and "counseling"—to controls, sanctions, and punishments. At the end of this process, we may see the situation predicted by the well-known geneticist Bentley Glass: "Unlimited access to state-regulated abortion in conjunction with the now-perfected techniques for detecting chromosomal anomalies . . . will free us of a certain percentage of all births with defects that today are still unmonitored. . . . In this future, no parents will have the right to burden society with a deformed or mentally incompetent child."[30]

The march of progress, which the advocates of the new technologies assign to the realm of individual decision, is thus viewed quite differently by the critics of this development: as a *process of socially reinforced inevitability*. What this means in concrete terms will be illustrated in the following section through two examples: the pill and the new methods of treating infertility.

2. The Pill: From the Possibility of Birth Control to Birth Control as a Duty?

The attempt to avoid unwanted pregnancies is almost as old as human history. But whereas most of the methods used in the past were not very reliable, in the second half of the twentieth century, methods of contraception are available that offer, for the first time, nearly 100 percent effectiveness. Thus it has become possible to avoid many forms of suffering: the constant fear of an unwanted pregnancy which burdened many marriages and love relationships in the past; or, in the case of an "accident," the problems of a single mother, a forced marriage, or an abortion.

Thus the advantages of the new procedures are obvious. And yet, after being initially welcomed with gratitude by many, a distinct change of mood has been noticeable in recent years. Typical is the attitude of women in the women's movement and women's studies. Although many of them initially saw the pill as a means of liberation from the constraints of biology and an extension of women's right to self-determination, a far more critical perspective has gained acceptance since that time. Gradually the "side effects" have become visible. First of all there are the health risks associated with the pill (from headaches to nausea to weight gain), and these risks are allocated to the woman alone, although sexuality affects both sexes. Moreover the changes in the realm of sexuality that have been initiated or speeded up by the pill also disproportionately burden women. That is, women have now become more easily available (because "there are no consequences"), whereas men are even more relieved of responsibility than they were before. This frequently intensifies the pressure of sexual expectations to the point of a "throw-away relationship" with the woman as object: "The liberating pill is becoming the compulsory pill. The positive possibility of having sexual intercourse without the fear of becoming pregnant is turning into the compulsion to have sexual intercourse."[31]

To be sure there was another reason why the pill became a focus of public attention: the sinking birth rate. The drastic drop in the birth rate since the midsixties, which has been much discussed by policy makers, scientists, and the general public, was quickly reduced by the media to a glib catch phrase and thus explained away as the "pill dent" made the headlines. Admittedly demographers countered that the pill is only a means to an end, and that means are employed only in situations where certain goals, wishes, and motivations already exist. But now it is becoming apparent that in fact new motivations arose during the sixties which were largely triggered by far-reaching changes in the lives of women.[32] If this interpretation is correct, then the pill did not cause the drop in the birth rate but did significantly contribute to it, simply because at the time it became available goals were also changing. In other words it was at just this point that ends and means coincided.

Furthermore the history of technology has shown at many points that a new technology is not neutral but carries within itself a whole program of social change. For example a new pressure to act arises through the pill: "In the past, if one did not want children one had to take some trouble with regard to contraceptives, but today in most cases one has to make a conscious decision not to use them if one does want children. As a rule, the decision process now runs in the other direction."[33] Under the impact of new options of birth control, there have also been changes in the attitudes, norms, and expectations in this area. We can safely assume that the development is proceeding more or less as follows. Because the pill very quickly hits the headlines in the mass media and provokes intense discussion among the general public, it also initiates a change of consciousness. Now it becomes obvious in even the most remote village that biology is no longer destiny and that instead there are options, namely, the decision to have or not to have a child. And in the interplay of questions, standpoints, and arguments exchanged in public discussion, there is a gradual shift in the "burden of proof." Imperceptibly a change in the prevailing social morality sets in. The ability to make decisions now becomes the *duty* to make a conscious decision. Or, to put an even finer point on it, the technology of contraception becomes the ideology of contraception.

> The new morality is conscious, rational, technologically guaranteed contraception. Its model is the enlightened modern human being, who deals responsibly with the act of conception. . . . In the age of unlimited possibilities of contraception, the person who does not use them almost becomes suspect. Contraception goes from being a necessary evil to being the duty of the enlightened citizen.[34]

"And what about the possibility of deciding *not* to control fertility?"[35] The advocates of the new reproductive engineering speak only of the possible choices that they open up. But on closer examination, it becomes evident that they always *close off* possible choices, too. For wherever contraception and planning become a normal life-pattern in society, those who do *not* plan, do *not* practice contraception, and do *not* adapt themselves to the standard image of the nuclear family soon become suspect (as being naive, backward, irrational, or whatever the labels may be). In short not practicing birth control now becomes a social stigma.

> This year, a woman I see every summer was pregnant. Again. It was her fourth baby in five years. I know she has money problems (and who wouldn't, with four children?) I know she's overworked and exhausted. . . . Four babies, I thought. For God's sake. And then we talked. She's a classic case: the woman who gets pregnant in spite of every method of contraception, even if she uses it absolutely correctly.

About this pregnancy, the doctor said, "Come on, I'll do an abortion immediately, and then you can go home no longer pregnant and forget the whole thing." She was tempted, sorely tempted. But no, she decided not to have an abortion. . . . She had made a decision, an unpopular reproductive decision, one that was not socially supported, even by her circle of friends. . . . While we, on the one hand, worry about the possibility of losing the right to a legal abortion—and this is a justified worry—we also have to defend the possibility of not having an abortion. . . .

Through the possibility of practicing birth control, the possibility of deciding to have a large family is simultaneously being made more difficult. North American society is entirely geared to small families, if it is geared to children at all. Everything—from cars and apartment size to the ideal picture-book family—is an argument for limiting fertility. Without good medical care, without child-care centers and appropriately sized apartments, children are a luxury—they're fine if one can afford them. . . . Must the choice not to be burdened by repeated pregnancies be paid for by losing the possibility of deciding to have a large family?[36]

There is also the fact that the birth-control methods offered are often applied in class-specific and group-specific ways, that is, they target certain population groups in a very selective way. And this intensifies their socially explosive aspect. This can be seen especially drastically in countries where different ethnic and cultural groups live, particularly when these groups have different degrees of status and power. The United States is a vivid example. Here thought was taken early on about measures to maintain the primacy of the indigenous white population group, while other groups—immigrants, Blacks, Indians—were to restrict their procreation.

Class prejudices and racism crept into the family-planning movement while it was still in its infancy. Among the advocates of family planning, the view became increasingly accepted that poor women, black women, and immigrant women had a "moral obligation to limit the number of their children." What was demanded as a "right" for the prosperous middle-class women was interpreted as a "duty" for the poor.[37]

3. New Methods of Treating Infertility: From Hope to Burden?

Numerous procedures for treating infertility have been developed in recent years. These include methods such as hormone treatment, which has now become routine gynecological practice, and in vitro insemination, which today is still regarded as exotic but may soon fall under the heading of standard procedure. But all of these methods, whether conventional or exotic, have in common the good purpose they serve. They are meant to help couples who long to have a child but have been unsuccessful so far. By the roundabout route of medical intervention, they are relieved of the burden of their suffering, and a happy family life can begin.

This, at any rate, is the picture painted by the pacemakers of this development. But is it accurate? To be able to judge, we must know more about the procedures being used than we learn from the success stories reported in the mass media. The following section will provide some supplementary information to round out the picture, so that on this basis we can then ask questions about the consequences for the people who are affected.

First of all there are the procedures that are part of the standard repertoire of treatments for infertility today: temperature-monitoring and hormone treatment. Even here sexuality is submitted to medical control in a very far-reaching way—in the "ideal case" probably entirely. It becomes at once a compulsory exercise and a competitive sport, to be performed strictly according to technical instructions (when to do it and when not, how often, and what position to use). Thus sexuality is regulated and disciplined, reduced to a purely biological activity. What is being lost along the way is the other, "superfluous" moments of sensuality, spontaneity, and feeling. Pleasure becomes frustration: under the pressure to function properly, both the inner self and the relationship with the partner suffer.[38] Here are two accounts of such experiences.

> The worst thing about infertility is making love according to plan. This takes away all spontaneity. I went through a phase where I wanted to be with him only on my fertile days. On the other days it seemed to be useless.

> At some point there came a phase when our sexuality went completely downhill. It really didn't matter very much any more. It was

a bit sticky and not very exciting; a bit tense. I had organized it completely.[39]

If we go on to the various procedures of higher medical technology, we find a number of other factors besides the regulation of sexuality. The procedures that are used are prolonged, time-consuming, expensive, and associated with considerable health risks and emotional strain. Here is a description of the individual stages of in vitro insemination.

It begins with hormone stimulation, which is "monitored via continuous laboratory tests." Once an increase in the hormones is established, the ovum is removed; in this procedure, "the abdomen (must be) cut open as it is in an operation." Because this always requires anesthesia, which has led to "lethal accidents" in the past, nowadays another procedure is frequently chosen, that is, the extraction of ova through the vagina. According to a precise schedule, that is, four hours later, "sperm is obtained through masturbation, which many men find embarrassing." After fertilization in a test tube, there follows a "quality check for embryos," "so that only those embryos are returned that at least appear to be regular." In this procedure, the embryo is examined under a microscope, and its "feasibility is graded on a scale of 1 to 5." If the embryo has successfully passed this "checklist for embryo quality," it is implanted into the woman's uterus. Further hormone tests follow, as well as, eventually, "the removal of hormone samples at short intervals," accompanied by constant tests. Nonetheless in many cases pregnancy does not result; or, because of a spontaneous abortion or a miscarriage, the wish for a child remains unfulfilled despite the pregnancy. Thus we must also be aware of "the hope and the disappointment, the physical and emotional pain, of thousands of women and men who believed that in being accepted into the reproduction program they had already nearly attained the fulfillment of their desires." And what if the pregnancy goes well? Even then "the mothers' troubles are not over: half of them will not be able to avoid having a Caesarean section. After this arduous series of treatments, the feeling is that a risky childbirth should not thwart final success." We can only add that thus a considerable health risk to the woman must be taken into account, for "even today, a Caesarean section (is) three to four times as dangerous as a natural birth."

Is this a negative picture painted by the enemies of progress? Not at all. It is only a sober report of what is happening here and now, described by those who know it best: the pioneers of in vitro insemination.[40]

The question remains of what success such a multitude of efforts has been; the statistics are sobering. Most of the couples who submit to treatments using the new technologies are not helped by this therapy to have a child. This is especially true of in vitro insemination, the target of so many desperate hopes. Here the success rate is still very low. According to official estimates, it is 10 percent to 15 percent, and critics point out that this percentage is probably considerably inflated.[41] But, to quote once again this method's advocates: given the present outlook for its (lack of) success, the rapid increase in the number of laboratories offering in vitro insemination may "prove to be devastating for couples affected by childlessness."[42]

For even in the cases where the treatments remain unsuccessful, they do not remain without any effects. The women and men who remain infertile—and they are in the majority—are not relieved of their burden of suffering by the medical intervention; on the contrary this burden is greatly increased. Moreover it is supplemented by what could be called "iatrogenic suffering," that is, the burden caused by the series of medical procedures and the continuous process of defining them as patients and people who are ill. In many cases self-image and self-awareness are damaged, life with the partner is put under pressure, and contacts with friends and acquaintances become less frequent. None of this is surprising: the time-consuming medical treatments leave less and less space for other interests and areas of life.[43] The couple focuses its thoughts, feelings, and actions on the child—who doesn't come. As Jacques Testart, one of the "fathers" of the first French test-tube baby, writes of his own experiences in medical practice:

> What would have happened to the couple's wish if such techniques did not exist? How often would this wish have been able—as it has done in the past—to follow the path of sublimation: the path of love for a child that has already been born, or even for a dog; or the wish could have been absorbed in reading, traveling, or artistic creation. . . . We don't want to conceal the marks of suffering left by the redirected wish. . . . But why should they always be contrasted only with the relief that science brings to a few lucky ones? Where, in this comparison, is the suffering of the other, more numerous, couples who remain infertile. . . . in whom the noise of progress re-awakened a wish that was about to be lulled to sleep, and who finally were forced, at the end of their physical and mental suffering, to experience the inevitable emptiness of failure. . . . We also have to talk . . . about the distress that is worsened by in vitro insemination or even created by it. . . . Something is morally rotten in the statistical expression of success.[44]

To this, one could reply that of course everyone has the option of walking away from the cycle of treatments. But this is much more difficult than it seems at first glance, if we look not only at the biological but also the social processes at work, and this difficulty is due precisely to medical developments. A "side effect" of this research is a redefinition of infertility and a lengthening of it on the time axis. If there are so many methods of treatment, then why not just try out the next one?

All of these new methods of treatment have also laid a new burden on those who are affected—the burden of always having to make an even greater effort. How many dangerous experimental medicines, how many months—or is it years—of compulsive temperature-taking and tortuous sex does one have to experience before being allowed to admit honorable defeat? When has a couple "tried everything," so that they can finally stop?[45]

Although infertility used to be one's allotted destiny, today it is becoming, in a certain sense, an "individual decision." Now those who give up before they have tried out the newest and latest method (an unending cycle) "have only themselves to blame." After all they could have gone on trying.

At what point is it simply and clearly not their fault, but something beyond their control—an inevitable destiny? At what point can they simply go on with their lives? If there is still another doctor they could try, another method of treatment, then the social role that infertility implies will always be regarded to a certain extent as having been freely chosen.... Is the fact that infertility is no longer regarded as inevitable but rather a "free decision" actually increasing people's freedom of decision and their possibilities for taking control?[46]

An important impetus is also being given to the new definition of infertility by those physicians who feel obliged to stress the particular significance of fertility in order to justify their expensive research. For this reason one of their standard arguments is the idea that motherhood has central importance in women's lives. This might be phrased as follows: "For many women, having a child of their own [is] not only their dearest wish but also their deepest biological need."[47] Or, to put a finer point on it:

In the last analysis, having a child is a woman's most essential task. This is what a woman lives for: to preserve the race, or in any case the species, of *homo sapiens* or *homo erectus*. This is her most essential task. Everything else that is added to it, a profession or whatever, is secondary. If a woman is not capable of this, then the whole essential purpose of her life remains unfulfilled.[48]

Thus the technology of reproduction becomes the ideology of reproduction. And what happens then to the women who don't manage to accomplish this "most essential task"? Don't they have to regard themselves as failures, incapable, inferior, and useless? The motto of one in vitro insemination clinic is, typically, "You're not a failure till you stop trying."[49] Surely many would rather risk trying yet another treatment than carry such a label for the rest of their lives. Thus medical interventions, which are after all supposed to help people, create new definitions—worse yet, new forms of *social stigma*, which make people more dependent and trap them in the cycle of being professional patients.

4 From the Wish for a Child to the Planned Child—In the Supermarket of Reproductive Engineering

The last chapter dealt with the kinds of inroads being made by reproductive engineering in the lives of women and men, their plans, expectations, and wishes. This chapter[1] focuses once again on reproductive engineering but from a different perspective. What interests me now is how these techniques are infiltrating the parent-child relationship. The basic idea, formulated at the outset as a thesis, is *the more these techniques become generally accepted, the more change there is in the "set of requirements" regarding parenthood*. And this is happening imperceptibly, in a multitude of separate steps, each of which nonetheless consistently builds upon those that went before. At the end of this development, there may be a radical reevaluation of everything we conceive of today as education—its private side as well as its public one. In setting forth this scenario, I want to proceed once again in two steps.

I will begin by briefly taking stock, starting with the past and then moving toward the present. Here I will take a look at the multitude of *demands and expectations that are placed on parents*, that is, the culturally prescribed outlay for the sake of children. This involves asking: How has this demand for education developed historically? Where are its roots, and what are the driving forces behind it? What is increasing parents' burden of responsibility? In this area there are reliable conclusions based on the research carried out to date.

41

In the second step I will be entering hazardous territory and deliberately launching a few speculations. Here I will first observe what is currently happening in the area of reproductive engineering and is already being implemented in clinics, laboratories, and research institutions. I will then examine the implicit consequences for parenthood. For the possibility is in the offing that the pressure of expectations on parents is not only growing quantitatively but also taking on qualitatively new forms—that the aim is *genetic quality control* of offspring. Against this background my provocative final question is Will our children's future be determined more by bioengineering than by education?

1. The Culturally Prescribed Outlay for Children

The Discovery of Childhood and Child-Rearing

In preindustrial society there was no child-rearing in the present-day sense, specially geared to the child's needs in terms of its age and development.[2] Not until the transition to modern society does the "discovery of childhood" (Ariès) begin. Soon it was associated with attempts to influence the child's development. The credo of the new attitude toward the child was that through proper care and rearing, parents could contribute significantly to the child's healthy development and even lay the foundation of its entire later destiny.

A look at the social history of the eighteenth and nineteenth centuries reveals that two circumstances primarily encouraged this new interest in child-rearing. For one thing this was the epoch of transition, in several phases, from a traditional class-determined society to an industrial society regulated by the laws of the market. Thus the individual's social position became more fluid and manipulable; it was no longer fixed at birth to the same extent as in earlier times. Education, in turn, gained increasing significance: now that social status was no longer simply inherited, there was a growing demand for knowledge and skills. Therefore educational efforts now focused on the child, concentrating on education and training so as to assert the individual's social status, safeguard it from decline, and even improve it if possible.

Second, the transition to modern society was accompanied by an ever-increasing faith in progress, aimed at the world's capacity to be mastered. In many fields experts arose who furthered the conquest of nature through theoretical and practical knowledge. Because of the progress made by medicine and later on by psychology, the nature of man also increasingly came to be viewed as fluid, open to influence, and capable of improvement. An obvious consequence was that a strong interest in the child was also awakened: a child is still at the beginning of its life; it is fluid and malleable. It provides an ideal field of activity where one can exert an influence—in accordance with the new worldview—to promote the developments one wants and prevent unwanted ones.

Against the background of these circumstances, the eighteenth and nineteenth centuries marked the beginning of the stage of deliberate influence (initially in the bourgeoisie and considerably later in the lower classes), in comparison to the merely basic care of earlier times. Initially this included efforts to protect the child from dangers to its health and harmful environmental conditions. Doctors advised proper infant nutrition and clothing, preventive medical care, and improved hygiene. Other forms of deliberate influence included attempts to guide the child's mental and moral development. Here an important role was played by the demand for education emanating from the philosophy of the Enlightenment: "Man can become man only through education. He is nothing other than what education makes of him" (Kant). The more these maxims consolidated into a cultural model, the more pedagogical tasks multiplied. Social learning, the child's language and schooling, its moral sense and spiritual welfare—all of these now became duties augmenting the work that had to be done for the children's sake. The educator Andreas Flitner writes: "The essential claim of the philosophy of the Enlightenment, with its respect for the human being as the subject of inalienable rights, and with its will to see in every person an individual, an independently thinking being capable of making decisions, is now granted to the child as well, at least prospectively: as a *task of the parents* to invest the child with such rights."[3] A new era began: the era of child-rearing as deliberate educational work.

As the recent history of pedagogy has brought out, these new models of education admittedly had a double root from the beginning. They are anchored not only in the claim to education that constitutes the ideal of the Enlightenment, but also in the compulsion exerted by the socially mobile society on the individual to safeguard his social status through education and training. Thus in this connection education acquires a double face: it is not merely stimulation but also, from early on, the pressure to succeed.[4]

Increasing Demands: "Optimal Starting-up Chances" for the Child

This claim to stimulation, which begins with the modern age, intensified continually as time went on. It gained additional weight in particular through various developments in the second half of the twentieth century. First of all there was further progress in medicine, psychology, and pedagogy that made the child seem increasingly malleable. For example physical handicaps (which at the turn of the century still had to be accepted as fate) became increasingly subject to treatment and correction. In psychology research took a new direction, emphasizing far more strongly than before the significance of the first years of life and even equating the failure to encourage a child with its loss of its opportunities to develop. At the same time a distinct rise in income brought the possibility of education, which used to be reserved for a small elite, within the reach of broad groups. Finally, at the political level, a promotion of education began that benefited previously disadvantaged groups. As a result of this and similar circumstances, the culturally determined pressure intensified. It became less and less permissible to accept the child as it is, with its physical and mental idiosyncrasies, and perhaps its faults as well. The child became the target of a multitude of efforts. As many faults as possible were to be corrected (no more squinting, stuttering, or bed-wetting) and as many aptitudes as possible were to be nurtured (a boom for piano lessons, language camps, tennis in summer, and skiing courses in winter). In books, newspapers, educators' advice—everywhere—the instructions sounded much the same: parents should do all they can to give the child "optimal starting-up chances." In sum this leads to the conclusion that, as a recent study of family development put it, the "norm of consciously responsible parenthood" is becoming increasingly accepted; in fact "the ethical and social responsibility of parents . . . (has) taken on a magnitude that could not have been historically foreseen."[5]

But why—this is the question that comes to mind—do parents go along with all this, anyway? Why don't they push aside all the pedagogic information that burdens them with ever more responsibilities and duties? The answer is that in our society there are many barriers that make it difficult to break out of the thicket of advice. In the first place parents are confronted on virtually every side by the commandment to provide the best possible education—from television to the newspapers, from advertising to the schools. And the refrain of this message is that ignoring the child's needs causes irreversible damage, and a lack of stimulation causes retarded development or even failure to achieve. Parents understand very well the significance of the phrase "failure to achieve," for "achievement" is a key category in our socially mobile society. Wherever the possibility and promise of social ascent exists, which always includes as its obverse the danger of social decline, the compulsion becomes ever keener to safeguard one's place in the social hierarchy through individual planning, efforts, and educational processes.

As an American women's magazine puts it, "Unstimulated time is a waste of baby time."[6] For the sake of varied stimulation, mothers (and fathers, once in a while) take the child to the circus and the zoo, attend a swimming course with the baby, organize parents' initiatives, and children's block parties. "Natural" childhood is in many respects a thing of the past; "childhood management" is beginning. And here, too, it's hard to refuse and will become even harder in the future. For these management activities do not originate, of course, in parental whim. The objective reason for them is that under the conditions of the mobile society, child-rearing and pedagogical enrichment are part of "status-maintenance work."[7] Wherever the compulsion to secure one's social status through individual effort dominates, it will necessarily find its way into the child's room as well. Child-rearing is caught between the hope of gaining social status and the fear of losing it.

"An example of this is a book of advice on child-rearing with the typical title *How to Raise a Brighter Child*. It begins with the sentences: "Here is help for parents who want to be sure that their child has every opportunity in life. For the road to success begins during the first six years of life, when the child's intellectual and emotional development depends on its parents to a very great extent, and when . . . the parents can significantly raise the child's level of usable intelligence through the type of child-rearing they provide."[8]

To sum up we can say that in highly industrialized society the physical care of the child has indeed become simpler in some respects, thanks to the mechanization of the household and prefabricated products such as baby food and disposable diapers. But, on the other hand, the discovery of childhood has increasingly led to the discovery of new issues. In Ariès's words, "Our world is veritably obsessed by the physical, moral, and sexual problems of childhood."[9] Thus many new tasks have been added at another level, as has been pointed out in a study of the sociology of the family: "Today the family labors under a *pressure to educate* which is historically unprecedented."[10] And this probably applies even more to the future. For the parents/mother, the child, who was once a gift of God and sometimes also an unwanted burden, is increasingly becoming "a difficult object of treatment."[11]

2. A New Parental Duty: The Perfect Child?

In recent decades, the task of the "optimization of starting-up chances" for the child was primarily interpreted as meaning that the essential factor was to provide the child with good schooling. Advertising for education in the sixties played a particularly important role here. The results are clearly documented in the statistics on education.[12] Fewer and fewer children are graduating from vocational schools. More and more are attending college-preparatory schools; still more and more are going to college.

And what will the future look like? In a society that is highly industrialized, complex, and linked up with international networks, and therefore requires a high level of achievement with regard to skills, such educational efforts will certainly continue to be necessary. But beyond that the responsibility parents are expected to bear may take on additional qualitatively new forms in the future. The triggering factor might be the new developments in medicine, biology, and genetics that are creating ever more possibilities for a precisely targeted "construction" of parenthood. "Programmed heredity"[13] is coming into reach, and the "tailor-made human being"[14] is becoming the vision of a possible future. The following section sketches out, in a deliberately speculative way, some possibilities that could be in store for us in this area.

New Questions, New Decisions, New Burdens of Responsibility

The all-important starting point is that reproductive engineering can be brought into play to fulfill the project of the modern age, the "optimization" of starting-up chances for the child. The following formulation, used recently at a convention of human geneticists and physicians in preventive medicine, could serve as a motto: "In our age, which focuses on achievement, even minor disorders and handicaps gain dramatic significance for development, integration, progress, and self-assertion."[15] By reversing these terms, one can deduce what is to be expected—in terms of these demands regarding health, fitness, and achievement—if technology keeps offering ever more possibilities for intervention. I would like cautiously to sketch in the scenario of the future in the form of a question: Will responsible parents still dare, in the future, to expect their child to bear the burden of a possible handicap? Won't they be obligated to do everything in their power to prevent any sort of impairment?

What this means can be defined further in more concrete terms. Parents who bow to these demands must, first of all, make use of the entire apparatus of prenatal diagnosis. This is not a small spectrum, and it increases daily. Genetic decoding of the human being is advancing rapidly, and the more aspects that can be diagnosed, the greater the probability that a defect will be discovered. If the finding is unfavorable in this respect, the parents will most probably decide to end the pregnancy. Wouldn't any other possibility mean that they want to burden the child with a life that would begin with inferior starting-up chances?

And that isn't all, by any means. If we think it through consistently, the parents' duty to care for the child must begin even earlier—at the point of conception. Responsible parents of the future must ask themselves whether their own "genetic material" is sufficient to the demands of the time or whether they should resort to a donated ovum and donated sperm—carefully selected, of course. The ethicist Reinhard Löw provocatively invokes the following vision: "In this brave new world, having one's own children means sending them on their journey through life with the irresponsible disadvantage of lower intelligence and a more modest appearance than the children who have been progressively conceived or combined in a test-tube. Already one can almost foresee the time when children sue their parents on the grounds of 'inadequate genetic inheritance.'"[16]

The Scenario of General Acceptance: The Beginnings

Accordingly the possibilities that may be in store for us through bio-engineering amount to the deliberate "quality control of offspring." Certainly such possibilities seem far off to us today. But what a general acceptance of them might look like can already be described. For as the history of technology has shown many times, the sequence of events from invention to general dissemination often takes place in similar steps.[17] And the development of reproductive engineering to date suggests that here similar patterns are emerging. The line from the present to the future might then proceed as follows.

At present only a few parents are beginning to make use of medical technology. These men and women are by no means aiming in every case to have a child that is as perfect as possible. On the contrary they have very different individual motives. First there is the most familiar case, which has been publicized sufficiently in newspaper headlines, of the infertile couples who now can have their longed-for child via medically assisted procreation. Then there are members of groups that the latest findings of science identify as "risk groups," that is, those in which the probability of a genetic defect in the child is increased. Then there are men who, before having themselves sterilized, deposit sperm in the sperm bank as a kind of insurance.[18] Similarly women can try the route of in vitro insemination in case they have themselves sterilized but later—for example, with a new partner—decide to have a child after all. And finally in modern society there is also a growing number of single people who want a child who fall back on the possibility of artificial insemination.[19] Varied though these motives are, they obviously lead in roughly the same direction: reproductive engineering. Many of those who decide in favor of it move into the next phase almost without noticing it. In the course of technical application and the possibilities that open up as a result, the wish for a child can quickly lead to the wish to influence the nature and characteristics of this child. Imperceptibly the road is being paved for a new form of "planned-child mentality."[20] This is not accidental but already programmed into the process for reproductive engineering makes selection possible, often even necessary. As Jeremy Rifkin provocatively puts it, the "inherent logic of this technology is eugenic."[21]

A vivid example of this is, first, the cases where the wish for a child is to be fulfilled by means of a sperm donor or a surrogate mother. When registering candidates, clinics and placement agencies make an initial screening. Admittedly the criteria are crude and more or less arbitrary, and they differ from one institution to another: from the individual's medical history and psychic stability (whatever the standards of measurement for this may be) to race, skin color, and height. An informational brochure from the Federal Ministry of Justice states that the aim of "this selection (is) to prevent the transmission of severe genetic defects by ascertaining the family history of the sperm donor."[22] A physician in Essen who specializes in artificial insemination in his practice includes in his criteria for selecting sperm donors: "No protruding ears or hooked noses, at least 1.75 cm tall, no 'flipped-out types,' a 'proper family background.'"[23] Another German doctor tries, in his own words, to take into account "character traits, appearance, and human appeal."[24]

At first glance such procedures are quite unobtrusive and harmless.

In the institutions' self-image, they are certainly not being used in the sense of eugenics and selective breeding; their purpose is only to rule out obvious defects or unacceptable characteristics in the client's interest. In this version of the situation, the crucial question, of course, remains unasked: What does "unacceptable" actually mean (e.g., a Black sperm donor for White people)? What is a "defect" (e.g., protruding ears)? Selection of this kind is no longer randomly distributed. It is by definition eugenic—not positive eugenics as selection of the best but negative eugenics as rejection of the "bad." And whatever/whoever is "bad" in this sense *also* depends essentially on social standards of measurement and value judgments.

Moreover the procedures of some institutions are so structured as to allow their clients—that is, people looking for a sperm donor/surrogate mother—to make their own selection from the possibile choices the institution offers them. Thus in the United States the standard procedure in some places is to provide clients with a catalog introducing all of the available sperm donors/surrogate mothers by means of a short description listing their physical, biographical, and social characteristics. From this catalog the clients may—no, must—make their selection. But if they have to choose, then why not make the "better" choice? If one had to choose between various articles, who would deliberately take the one he/she likes less? So, too, in this situation: since a choice has to be made in any case, it seems reasonable to assume that each individual will choose according to his/her own ideal image, so as to steer the genetic roulette wheel toward certain characteristics. Accordingly some "customers" put their money on intelligence, others on musical aptitude, and others on blonde curls or athletic achievement.

By way of illustration, here is an example from a surrogate-mother catalog:

Martha F., *Address*: Escondido, Ca. *Pregnant*: No. *Status*: Divorced. *Employer*: County . . . *Birth Date*: 6–11–48. *Height*: 5'2". *Weight*: 110. *Hair*: Blond. *Racial Origins*: Caucasian. *Children*: Kammy, Age 14, and Crissy, Age 10. *Medical*: Normal delivery both children . . . no surgery or other problems. Medical release and detailed medical history completed. *Could begin*: Immediately. *Insurance*: Greater San Diego. *Expenses Anticipated*: $20,000. *Photographs*: Available. *Contact*: By mail forwarding. *Comments*: My older child is a mentally gifted child. (Interested in planned or surrogate parenting.)[25]

But here, too, the wishes, individual though they seem to be, are by no means randomly distributed. Rather they reflect the hierarchy of social values. Wolfgang van den Daele, a member of the German Bundestag's Commission on Genetic Technology, has described this connection as follows:

> Certainly the "ideals" of health, beauty, intelligence, and appropriate behavior are to a large extent socially defined. But they do not need to be forced upon us. We have internalized them, and we reproduce them as our own needs. This is why, if there are technologies which allow us and our children to approach these ideals, we will also *want* to use them. The less probable it is that eugenics and the breeding of human beings will be forced upon us by the state, all the more likely is it that "consumer choice," in the guise of self-determination, may become the means by which it becomes generally acceptable.[26]

A striking example is the case of an Austrian aristocrat whose wife is infertile. Now he hopes to have six sons by means of an American surrogate mother. He has already had three by means of this procedure, and the fourth one is in production at the moment. Every time, the doctor takes the trouble to separate out the "female" sperm in a centrifuge so that it will once again be a son.[27] Similar cases of the combination of surrogate motherhood with gender selection have also been documented elsewhere.[28]

A further example of the selection processes being set into motion by reproductive engineering is prenatal diagnosis. Here the current legal situation in the Federal Republic of Germany is as follows: an abortion is permitted only if the woman cannot be expected to continue the pregnancy because there is the risk of severe and irremediable damage to the child. But the actual social situation admittedly looks quite different. Even today the social practice of prenatal selection is overrunning the normative limits that criminal law tries to establish. Increasingly abortions are being carried out even in cases where a less severe or a treatable (e.g., operable) defect is expected. Here is another comment on this by van den Daele:

> The women (or parents) who are affected by the results of prenatal diagnosis often react with an "all or nothing" attitude. As a rule, abortion is chosen even if there is only a certain risk of illness—that is, the probability that a healthy foetus is being killed is relatively high—or if it is impossible to decide whether an expected defect will be severe or minor. . . . Even the diagnosis of chromosome anomalies (such as XYY), which are almost certainly clinically insignificant, is taken as a reason to abort the foetus, "just in case."[29]

Already in some American surrogate-mother contracts, the surrogate mother commits herself to submit to prenatal diagnosis and have an abortion if the results are unfavorable.[30] In the combination of technologies and their promises, a "quality product" is envisaged. The wish for a child is becoming the made-to-measure planned child.

Momentum, or, The Carousel Whirls On

These are not visions of a distant future but developments we are already experiencing today. From them we can gauge how reproductive engineering is becoming a gateway for procedures that de facto amount to quality control of potential offspring. As we have just shown, this is happening even though the clients/patients initially come because of entirely different motivations. But once they have arrived in the wonderland of technology, their wishes suddenly develop a power of their own. They become independent. This is not an individual fit of self-indulgence but something inherent in the inner logic of the process—a tendency that is very welcome to some geneticists.

H. J. Müller, who received a Nobel Prize for his work on the effects of radiation on genes, concluded in 1959 that thousands of women had already had themselves artificially inseminated. Infertility, wrote Müller, provides "an excellent opportunity to strike a blow for positive eugenics, since the couples in question are almost always open, under the given circumstances, if you suggest to them that they make a virtue out of their necessity and have as many gifted children as possible."[31]

Thus do new means generate new ends. The American sociologist Joan Rothschild writes, "The new technology, in terms of its effects, helps to set new standards of human perfection and human defects. . . . Within the space of two decades, scientific and technical knowledge has changed the realm of possibility so greatly that our concern about having a *healthy* child has been replaced by the pressure to have a *perfect* child."[32]

Even today the doctors in genetic-counseling centers are being confronted with the wish for a prenatal genetic diagnosis even in cases where this diagnosis—according to the catalog of medically defined risks (e.g., the age of the mother)—is by no means indicated. For example nowadays even some younger women want amniocentesis, which is routinely offered to older women. This can be interpreted as an irrational reaction expressing individual personality traits, for example, a strong tendency to anxiety. But this kind of interpretation overlooks the social genesis of such expectations: they are systematically produced by the offerings of technology. The more people know about prenatal diagnosis, the more will illnesses and predispositions to illness move into public awareness—thus generating new expectations regarding intervention. "Inevitably, things which are technically possible and feasible that are reported on in the public sphere, from television to the playground, will be increasingly used and demanded. . . . The trend toward more prenatal diagnostics, through the interplay of supply and demand and the expansion of possible methods, can be clearly seen everywhere," says human geneticist Traute Schroeder-Kurth.[33] And Jacques Testart, himself one of the pioneers of in vitro insemination, describes vividly how medical technology brings about an inflation of desires. In a sarcastic vision of the future, he describes how in vitro insemination will lead one day to the offer of "ova à la carte." Parents will then have the freedom to choose from an assortment "like that of a pet shop: hair color, leg length, ear shape and health certificate."[34]

But once a planned-child mentality of the kind described above has arisen, it can expand rapidly. For the parents who first acquire this mentality create standards of measurement which then in effect apply not only to their children but to *all* children. Thus other parents will join them on the road to bioengineering. They will follow suit because they fear that their child may otherwise not be able to keep up in the competition of an achievement-oriented society. And at a later stage, the state may also begin to resort to guidance mechanisms to protect the children with "less responsible" parents from starting-up disadvantages that could have lifelong consequences. The individual stages of this scenario of general acceptance are sketched out by van den Daele:

> Scientific progress will create new possibilities for influencing the genetic make-up of human beings. And if correlations between genes and I.Q. should be discovered, there will certainly be parents who select or constructively improve their children's genetic substance in order to equip them with the best starting-up conditions for the competitive struggle of their later lives. But once this has begun, who can then afford to hang back? Given that an appropriate technology exists, the standardization of genetic intelligence may become a veritable parental duty and a social norm. Finally, at some point it may no longer appear implausible for the state to enforce a certain minimum of "genetic care" for the sake of the child's well-being, more or less as it does today with respect to schooling.[35]

We see here that what initially begins in small steps, limited to groups that are the exception, can quickly develop a momentum of its own which overwhelms our previous patterns of living. This is also known from the history of technology. Wherever new possibilities for action are opened up, standards of behavior also begin to change. What "consciously responsible parenthood" might then look like has been described by Jeremy Rifkin as follows:

> The social pressure to conform to the logic of genetic engineering will be enormous. . . . In the future, every set of parents will have to decide whether it wants to try its luck in the traditional genetic lottery or to program certain characteristics into or out of the child-to-be. If the parents prefer to behave traditionally and let genetic accident determine their child's biological destiny, they could get into great difficulties if a fateful mischance, genetically speaking, should result.[36]

The Ideological Effect

We have still not reached that point. Moreover, as far as the control of human intelligence is concerned, the technical achievements of modern genetics are modest and will probably remain so in the foreseeable future. (It is not yet possible to breed the perfect child, only to "prevent" so-called defective children.) But the discussion of future research perspectives, as it is being carried on here and now, is in itself giving new impetus to the connection between intelligence and genetics. And this connection, as van den Daele has once again pointed out, is even now potentially devastating.[37] The concept of "innate destiny" is thus gaining new adherents. It is becoming increasingly plausible and legitimate to look for the conditions of human development more in man's biological nature and less in his social and cultural environment. Against this background demands for equal opportunities in the educational system lose their power to gain general acceptance at the policy level, because opponents can always point to the determining power of the gene. Whereas in the past social reforms seemed necessary, in the future parents will be able to put their money on the "genetic improvement" of their offspring.

In this manner eugenics, too, could once again become respectable—in a new guise, to be sure, as a sort of "gentle eugenics." In this connection ideas about "superior" and "inferior" groups do still exist in society (e.g., sorted with regard to race, class, and gender). But with the new technologies, the ideal models of what is signified by "better" will be put into a less threatening framework. Joan Rothschild describes this process of transformation:

The more accessible these reproductive technologies become to the members of the middle class and the upper middle class—that is, the more the parents are assured that they will have an approximately "perfect" child—the more will the parents' genetic substance become the criterion for gauging the ideal. . . . Ensuring "better" offspring now means selection from the genetic substance of the parents or, conversely, selective rejection (or perhaps perfecting it in the future). The standards applied seem to be private ones. Whether one hands down better or worse genetic material to one's offspring becomes a question of personal responsibility and decision.[38]

Thus the technologies seem neutral. They threaten no one with destruction and annihilation. They don't aim to help the "members of the master race" to world victory. They don't disturb one's good conscience. And their attractiveness grows all the more freely. Their promise, their message, whispers softly: Is it wrong to want healthy babies?[39]

3. An Inexorable Development?

So far, it has become evident how a certain momentum is being generated, in the course of which gradual habituation to the offerings of reproductive engineering can begin. And what is happening here and now in many clinics, laboratories, and research institutions is already a part of this normalization process. Thus there is every reason to assume that further movement toward popular acceptance is an inevitable and inexorable process that can no longer be slowed down in any way.

And yet, when we look at it in fundamental terms, this assumption is false. For the popular acceptance of technologies is not a process that occurs according to natural laws; it is a *social* process. Its course is directed not by preexisting determinants but by social, political, and economic conditions. It is dependent on power relations and group interests, market shares and career chances, political priorities, legal regulations, and private decisions. Thus it is in principle open, stoppable, controllable. It can be braked when the opportunity offers itself.

So there is still an "escape hatch" for eluding this momentum. The question is, of course, whether we will find it.

The Ineffectiveness of Previous Controls

In any case at present the course is set in another direction. Bioengineering is invading ever widening fields; success stories are coming in thick and fast; the newest research results are constantly being used to open up new possibilities of implementation. Because of the speed with which all this is happening, the process of gaining public acceptance is practically uncontrolled: medicine is becoming "subpolitics."[40] A "forced march toward the future" is the result.[41]

To be sure there are numerous committees whose work specifically consists of drafting binding regulations concerning the questions and problems raised by bioengineering. For example there are commissions appointed by parliaments, statements made by churches and parties, and debates at conventions of physicians and jurists. But if we look more closely at their results, it becomes obvious that so far the processes of democratic monitoring have been insufficient and inefficient at key points. Van den Daele's comment on the final report of the Commission of Inquiry on "Chances and Risks of Genetic Technology" is symptomatic: "What the majority of the commission presents has the characteristics of a gentlemen's agreement with development, which cannot be reversed in any case, but which they do not basically call into question either."[42] Moreover such commissions face immense and completely novel questions which at no point fit the schema of specific scientific disciplines. Thus the struggle to find answers is a long one. Then, too, it takes time to negotiate draft bills, make changes, reformulate them, and toss them back and forth in jurisdictional wrangling between different departments. Years elapse in the process. But the pioneers of reproductive technology are not waiting until the state of affairs and the legal situation are clarified. In their laboratories they have been inseminating and creating life for years—in vivo, in vitro, homologously, and heterologously.

But the Consensus on Progress Is Starting to Crumble

Thus the natural sciences have de facto taken on a pioneering role in the process of social change. For a long time this pioneering role was legitimated by the belief in progress that characterized modern society. But in recent years a crucial change has begun at just this point. As the current discussions of social policy show clearly enough, this consensus on progress is beginning to crumble at many levels and within a growing number of groups. Criticism is increasing and this applies in particular to the interventions in the realm of human nature that are being initiated by bioengineering. Here a polarization of standpoints has clearly taken place, and it is striking that new types of coalitions are arising that no longer follow the usual political patterns. Thus in recent years a rainbow coalition of men and women scientists from a

wide variety of disciplines has developed—scientists who view biotechnology from different frames of reference but are unanimously critical
as they point out the unplanned and unwanted side effects of technical
intervention.

In this connection it is very significant that the reproductive technologies are also sparking controversy within the medical field. Some
doctors argue vehemently in their favor; others warn of their risks, and
some initially participated in this development but are now calling for
a retreat. One of these is Jacques Testart, the "father" of the first French
test-tube baby, who has since then pleaded "for a logic of non-invention,
for an ethic of non-research."[43] Similarly the physician Manfred Stauber,
after many years of clinical experience, is now demanding that those
involved in heterologous insemination take "a pause for thought and
even 'a step backwards'" so as to postpone the "plunge into the abyss
of an endless and uncontrolled reproductive medicine."[44] Equally drastic
is the verdict of Peter Petersen of the Medical College of Hannover, a
former member of the interministry working group on in vitro
insemination, genome analysis, and genetic therapy: "Our current level
of awareness in science and practical life can not comprehend the holistic
reality of test-tube insemination. The doctors who are working on test-
tube insemination don't know what they're doing."[45]

From this viewpoint we are in a situation where the new procedures
of bioengineering are already the focus of intense controversy but at
the same time are gaining widespread popular acceptance without en-
countering any restraints. This is an alarm signal for the political sphere.
Action is urgently needed to breach the monolithic dominance that the
natural sciences—or rather, some of its practitioners—have established
in the creation of facts and new standards for human life. Obviously
the viewpoint of the natural sciences alone is inadequate to this task.
What we need is a comprehensive discussion of the societal, social-
policy, and political consequences that are in store for us as a result of
these technologies.

Bioengineering Instead of Education?

In this chapter I have presented some reflections about the ways in
which reproductive engineering may place a new burden of responsibility
on parents. To sum it up in a phrase, "genetic quality control" may
become the task of the future. It is clear that at present these can only
be speculations, and it is equally clear that there are advocates of other
viewpoints who find such speculations exaggerated and excessive. Therefore

I would like to conclude by reminding the reader once again of two powerful motivations that are driving forces behind the trend toward "quality control" of offspring. First there is the fact that in the socially mobile society parents are under considerable pressure to do all they can to give their child optimal starting-up chances. In other words the demand for "quality control" already exists. Second we know from the history of technology that a new technology often helps in itself to create further demand. With each new promise the wishes grow.

If the thrust of this development is not checked—soon, before it's too late—then a change may take place in the paradigms of education and professional training in the future. To reduce it to a trenchant formula, it becomes bioengineering instead of education. The project of the Enlightenment, which came into being at the outset of the modern age and helped to determine its design for the future, would then clearly end with the dialectic of the Enlightenment. But surely it is still possible to find other pathways. I hope that the doubters are proved right in the end and that my speculations about the future do not come to pass. Perhaps this will become possible when we realize what new pressures and burdens of responsibility would otherwise be in store for parents and when we therefore set limits at the political level. In this spirit, I hope my ideas constitute a "self-destroying prophecy" and that they will contribute to their own refutation.

5 Pathways of Normalization and Constructions of Acceptance

Let's take stock of the present situation. On the one hand we see that the consensus on progress is starting to crumble. Whether the new biotechnologies will provide the opportunities hoped for by some or the risks stressed by others is being fiercely debated today in the political sphere. But we are also seeing that the development and application of these technologies is proceeding at a pace that leaves no time to think things through as they happen, let alone in advance—only *afterward*. Given this situation, urgent questions arise for policy, for society, for all of us: How can we escape the dictates of the feasible? How can we restore the openness of the future?

In this connection, Bernd Guggenberger writes:

The important thing to rediscover is this: that the rules of our civilization—our life, our activities, our dealings with nature—are the rules of the game we ourselves have laid down, so they can be changed . . . they are therefore not natural laws which impose upon us industrial civilization's imperatives of growth, progress and risk at any price; rather, they are self-imposed goals for action.

He continues: "The 'pressures inherent in the situation' that we allude to are nearly always pressures caused by people and their interests, which in turn impose continually expanding pressures on the thoughts and actions of these and many other people."[1]

Perhaps it would help if we looked at these inherent pressures once again more closely, this time in the spirit of the motto: *"Pressures inherent in the situation" do not come out of the blue. They are created by people.* In other words only the person who is familiar with the pressures inherent in a situation can escape them. So, what is the driving force behind research, application, and use? Is this development proceeding of its own accord, or, if not, *who* is the driving force, and what interests are involved?

Looking at the situation from this perspective, we first notice certain characteristics of the working conditions in science and research that open up paths of normalization. But we also find precise strategies that are used by the pioneers of the new technologies. The momentum that often arises in the general acceptance of technologies, which has been described above, is not a process in which agreement comes automatically, as though their dissemination occurred more or less as a matter of course. This is certainly not the case: the ball has to be pushed to make it roll. And the law of economics holds here, just as it does in other fields: markets have to be created; receptivity has to be built up; resistance dispelled; and coalitions forged. And the more questions and doubts surface, the more intensely this has to be done.

1. The Abstractness of the Viewpoint

One of our traditional expectations regarding the doctors' art of healing is that they will serve the human being as a totality that includes body and soul. But what is it, this unity of body and soul? What is happening to it? How can it be demonstrated? The procedures of sophisticated medical technology reduce complex processes, for example, the crisis of being childless, to a defective Fallopian tube or another defect that can be localized, and then try to circumvent this defect. "But in this process the broader and deeper context of this defect is viewed as merely marginal or wholly insignificant—such as the fact that the individual having this defect is a human being with a biography."[2] In many cases the result is that the crisis of infertility intensifies in the course of the treatment.[3] Here it becomes evident that the ever-increasing specialization is a central "production factor" in the genesis of so-called side effects.

> The *higher* the degree of specialization, the *broader* is the extent, number, and incalculability of the side effects of scientific and technical activities. What *arises* as a result of specialization is not only the "invisibility" and the "secondary nature" of the "invisible side effects." Together with specialization, the probability increases that local solutions are invented and implemented whose intended main effects are continually thwarted by their unintended side effects. . . .

Thus the . . . structure giving rise to "the pressures inherent in the situation" and "the momentum of the situation" is essentially the model of overspecialized application of knowledge with its limitations, its understanding of methods and theory, its climbing of the career ladder, and so on. This specialization of labor, carried to an extreme, produces it all: the side effects, their unpredictability, and the actual facts which make this "destiny" seem inevitable.[4]

And there is certainly no place for the human being in his/her wholeness under the microscope in the laboratories of molecular biology. There is room only for his component parts: the human being dissolves into the mosaic of his cells, chromosomes, and genes. In this abstractness the question of what human life is, what the special nature of man consists of, vanishes. We no longer see a living being but, instead, the thousands of strands of genetic information that represent the blueprint of the living being. This immaterial approach, which characterizes molecular reductionism, has enormous and far-reaching consequences. Palpable limits and boundaries disappear. Taboos that arise from the direct experience of living beings dissolve. The "visible threshold"[5] for experiments and intentional alterations is sinking. So why not do research using human embryos if this can help open up new pathways for treating infertility? Why not examine and sort out embryos at an early stage of their development, that is, not implant them at all if they show signs of damage? "The escalation of 'progress' is possible only because more and more experts who are cut off from one another feel responsible for only that part of the structure of scientific knowledge which has been assigned to them."[6] This is the opinion of Jacques Testart, who has had sufficient experience of this escalation in his own research.

Human genetics is a form of man's encounter with himself, with the project of his history. In the course of science, he has designed himself as a mechanism, and is now discovering his center as a formula, as a hybrid relationship between chemical substances and biological cell structures. In this process, a human essence has simply not turned up, not sprung forth, even with the best will in the world. Thus it is not *possible* to violate it either. In the abstractness of the laboratory—in dealings with the chemical banality of substances, which in their abstractness contrast with the thought of a living being, just like the formulas that express and yet do not express this living being—in this abstractness and because of it, the boundaries between death and life can be redrawn arbitrarily, i.e. in the absence of experience, nominalistically. Nothing hurts, nothing responds, nothing puts up a fight. . . . Here, nobody *can* know what he's doing anymore. And thus, gradually, anything can happen.[7]

2. The Freedom of Science and the Action of Thoughts

The definition of a technology as "experimental" is a social instrument often used in the natural sciences and technology in order to shield innovations—in their earliest and thus most vulnerable stages—from the attention and monitoring of the general public. This process of differentiating research thus has the effect of "culturally dedramatizing" it as a way of countering potential resistance.

> The process of differentiating research expands the realm of what may be legitimately regarded in society from a merely technical standpoint. In research, only the feasibility of technical possibilities is discussed; no decisions are made about their implementation. This limitation offers not only the advantages of specialization, it also relieves the genesis of technologies from the normative controls and selective conditions set by specific contexts of use. Through these circumstances, the differentiation of research is becoming an innovation that accellerates innovation.[8]

The postulate of the freedom of science arose in epochs during which it was important to create space for science against bans on knowledge imposed by the church and intrusions by the ruling powers. But since that time the range of action open to science has changed drastically. The time is long past when the researcher sat alone in his ivory tower, dedicated only to his ideas, surrounded by only a few test tubes—nowadays research in the natural sciences is a large-scale enterprise requiring expensive equipment and extensive funding, intertwined with many investments and interests. For these reasons it is not a mere bead game with no social consequences. The generation of technical possibilities does not remain in the realm of pure thought; it opens new realms of action and thus is two things in one, the "action of thoughts."[9] Under these circumstances there is often not much more left for Parliament and the parties than the vague prospect of coming to terms, later on, with a development that has been initiated somewhere in workshops or laboratories.

3. Strategies of Legitimation

Gaining general acceptance for new technologies is also a matter of politics. Acceptance has to be created, and to this end certain strategies of legitimation are used that are similar to those used in other areas of politics. Let me name a few typical examples from the broad field of possibilities.

The cliché "we've always . . ." is popular and frequently used in connection with the new technologies—that is, recourse to the argument that the new technology is actually not new at all, but in every essential aspect the equivalent of those already being applied and thus also justified. For example specialists in industrial medicine tend to view genetic tests to determine inborn susceptibility to work-connected risks as not being qualitatively new in comparison to the blood tests or urinalysis used until now. Similarly genetic engineers argue that in principle all they are doing is what classical breeding has always done only in a better and more controlled way.[10] Resistance to the project of genome analysis is countered by pointing to old Gregor Mendel.[11] And specialists in reproductive medicine begin an article on "In Vitro Insemination as Substitution Therapy" with a reference to the Middle Ages and antiquity:

> As long as the art of healing has been practiced, there have been attempts to restore to health an impaired capacity to reproduce. Thus reproductive medicine is not new, even though its methods and emphases have changed. In the Middle Ages, official medicine was characterized by the skill of Jewish and Arabian physicians, who generally practised in royal courts. Monasteries were the main repositories of medical knowledge. But gynecology and obstetrics were neglected because of the taboos that Christianity imposed on male physicians, although these skills had been taught and practiced in pre-Christian Rome, for example, by highly trained midwives. In Germany as well, women helped one another during childbirth. The medical practice of these midwives included methods of contraception and abortion as well as fertility treatments and love potions.[12]

At this point the lay person might well ask why, then, the drama of the new is extolled on other occasions: "Since the beginning of the Sixties, intrauterine space has been explored in a variety of formerly unimaginable ways, and at an extraordinarily rapid pace."[13] What a breakthrough! This sounds like the discovery of Africa or the exploration of the Wild West.

Why, then, do some people talk about business as usual, everything normal, nothing new under the sun? The logic of this approach is obvious: if something already existed yesterday (in the Middle Ages or antiquity), then it can't be wrong today—so there's no cause for concern, anxiety, or moral scruples. The introduction of the new takes place behind this protective shield: all of this has "in principle" already been practiced and recognized as worthwhile for a long time. As yet unknown risks and dangers are compared out of existence, minimized, redefined. If the forefather of genetic engineering was Gregor Mendel—just think of those nice red and white flowers—why should its further development in the areas of data protection, labor law, and insurance law raise enormous problems? Thus the cliché of "we've always . . ." is also used as a strategy of "cultural dedramatization": "Old technology has traditional legitimacy. And initiators of new technology usually seek a connection with this legitimacy by pointing out that the new aspects go only a negligible distance beyond what is familiar and already accepted."[14] To put it provocatively, "The future is wrapped in the past in order to bring it across the sensibility thresholds of the general public in the wrapping paper of the seemingly familiar, *invisibly* and without legitimation. At the customs office this kind of thing is called smuggling."[15]

Another way of creating consent follows the basic formula "for a good cause." Here the strategy is to associate the new technology with generally accepted values and make it unassailable through this alliance, for example, for health and the desire for children, for self-determination and freedom, for human dignity and progress, for research, knowledge, and truth, or, in more concrete terms, genetic engineering in the struggle against cancer, AIDS, etc. The pioneers of in vitro insemination declare enlightenment to be their ally. In contrast to previous centuries of fatalism, "the enlightened human being is not willing to submit to suffering for which there is a medical remedy."[16] Those who contest this expose themselves to the suspicion that they wish to obstruct the good. "Technical possibilities not only annul existing values, they also create new ones. By expanding the options for implementing the established values of a culture, technologies redefine these values' area of relevance . . . At the same time . . . the technical possibilities are thus integrated into the area protected by these values."[17] A dilemma is being created. Those who are against it are thereby advocating women's suffering due to their unfulfilled desire for children or children's dying of cancer. And the moral is always the same: a good end justifies the means.

This is the legitimating dimension. If society guarantees rights that correspond to the values, then the rights also extend to the utilization of the technologies. For example, from the constitutionally guaranteed protection of health, the freedom to start a family, or simply the general principle of self-determination, according to which everyone must be allowed to decide for himself how to structure his life and how to deal with his own body without outside interference— from these rights we deduce individual claims to also be allowed to use the available technical possibilities. Such claims are legitimate and can be contravened by regulatory policies only at the cost of an increased burden of proof and argumentation. . . . Social values are not only a counterbalance to the dynamics of technology, they are also its vehicle. The inclusion of technical possibilities in the area protected by individual claims shields them from political intrusion.[18]

The fact that the formula "for a good cause" is presented here as a strategy of argumentation meant to win assent certainly does not mean that the connection with generally accepted values is merely a fiction or a pure invention. Of course there are plenty of examples of ways these technologies can be applied, for example, to prevent illness and alleviate pain. So this connection in fact exists; indeed the strategy of legitimation draws its power from just this circumstance—that the connection is in many cases obvious.

But conversely the connection with values is in many other cases ambiguous, controversial in terms of its scientific basis, or not open to monitoring by laymen. Often the basis of legitimation is stretched and extended through analogies and comparisons that progress from certain to uncertain cases. Often no one knows exactly where the boundary actually lies. The "principle of hope" is an aspect of legitimation, but it is also the driving force of research—and how can we distinguish between the two? (Who can nowadays say with certainty whether genetic engineering can or cannot offer a therapy in the struggle against AIDS?) And this blurred line of demarcation, in turn, helps to extend the radius used for value judgments and legitimation.

As Ivan Illich provocatively phrased it: "The new specialists . . . like to go about their business in the name of love."[19] The new reproductive technologies are already associated with the piece of advice that it's better to space the births of one's children precisely, because too close a sequence of births endangers the children's health: "It has been proved that children who are spaced too closely together tend to have more genetic defects. Such risks can be eliminated if the wife has an ovum removed while she is still young, then has it inseminated in the laboratory with the husband's sperm and frozen until it is needed a few years later."[20] And when the practice of surrogate motherhood got caught in the crossfire of public questioning, it was counterargued that surrogate motherhood, if carefully planned, can eliminate many risk factors that occur in natural conception and "could be fatal to the child—unwanted pregnancy, poor parental health (emotional as well as physical), drug use."[21] From the development to date, we can deduce at least this much with certainty: generally accepted values—including, above all, medical aims—have thus far opened up still legitimate areas of application to the further technologization of the human being. So why not use embryos as storehouses for replacement parts, if this promises to further the aim of healing?

According to two pioneers of reproductive engineering,

> Embryos cultivated beyond the blastocyte stage can become "storehouses of replacement part" for differentiated human tissue. In cases where bone marrow, brain cells or liver cells are in danger of destruction due to illness, we could have secure reserves to fall back on. We are already transplanting hearts, kidneys, and livers after removing them from the recently deceased. The immune system can ruin an organ transplant but not the transfer of embryonic cells. Ova and sperm cells are available. But we resist this type of treatment with embryonic cells. Why, actually, do we react to this with such sensitivity, if embryonic life can save human life, for example in cancer therapy?[22]

And here, as in other areas, the motto is nothing succeeds like success. So it is an entirely normal procedure for even the pioneers of bioengineering to proudly trot out their success stories. Best of all is a vivid image: color photographs of happy mothers and cute babies.

One problem, however, is the fact that in many cases this success does *not* occur. This is especially true of in vitro insemination, which triggers so many emotions and conflicts. This is a blemish that leads to further criticism. Therefore the pioneers in this field have developed many ways of "creatively structuring" the statistics and thus increasing the success rate substantially, at least on paper. (An especially vivid example comes from a large American university which has a success rate of 25 percent, according to its own figures, although it does not yet have the birth of a single child to show for its efforts.[23]) This is popularly known as "cooking the books."

Complaints about such practices—which amount to the construction of "Potemkin villages"—are appearing nowadays even in renowned professional journals. But because many of the in vitro insemination centers operate according to the laws of the market, they are obliged to promote the attractiveness of what they have to offer if they want to safeguard customer demand. Medical researcher Urban Wiesing writes,

> We have to conclude that tinkering with the results is an indication of the medical profession's self-interest regarding this method, because the participating couples gain no benefit whatsoever from retouched statistics. On the contrary, they are being deceived. . . . It's less a matter . . . of the patients' well-being than of . . . especially good results within the group of competing centers, and of the general public's acceptance of the method. For the medical profession, it's a matter of the unchecked development of its own technology.[24]

The old motto "the end justifies the means" also applies to Potemkin villages. The "hope business" is booming.[25]

But it would be too easy only to look at the advertising experts who market in vitro insemination. The best marketing strategies fall flat if there is no demand, or at least partial willingness, among the potential customers. This applies here as well. There are many indications that the general public often has divided motives. Many men and women regard reproductive and genetic engineering with unease or even reject it "in principle," but when the same men and women think about their own lives in concrete terms, the particulars of their own situations, then hopes stir and barriers fall. Each one of them knows, of course, that he himself acts responsibly, isn't one of the abusers, is capable of recognizing limits. For example, probably most of the women who know German history and have heard of euthanasia reject any type of eugenics. But in their own lives, many of them will nonetheless welcome prenatal diagnosis as a procedure that "balances out" the risk due to ageing and takes away their anxiety about a possibly handicapped child. (And they have good reason to do so. In terms of our working and living conditions, women who have a healthy child are already handicapped enough.) Other women who regard medical technology and "professional patients" with unease may still be tempted to try out in vitro insemination for themselves—for example, if they have had themselves sterilized but then want to have a child after all with a new partner. Another example is surrogate motherhood. Condemnation of surrogate motherhood cuts across almost all social groups, whatever their political orientation. Yet "despite this nearly unanimous rejection, business is booming for the agencies"[26] (at least wherever they are permitted, and elsewhere perhaps through darker channels). What is happening here is a split between "public" and "private" morality; what's more this split exists within the individual, who acts according to the compass of his own needs. This is no surprise to psychologists: the mechanisms of need-oriented perception and dissonance reduction function here just as they do in other areas. This characteristic ambivalence in turn helps to break down taboos and extend the radius of what is socially accepted. The process of gaining public acceptance follows a pattern that might be called the *"bridgehead* strategy":

> The question remains: "Why do new technologies win general acceptance even if they encounter resistance, i.e. if their integration is accompanied by conflicts?" This question points to the composition of the "receiving" social system into which the new technology is to be introduced. The crucial mechanism here is the fact that the new technology begins, at the moment it is socially perceived, to render formerly valid structures of expectation and behavior arbitrary and to open up new horizons of expectation.... The social organization that is required for the conception and implementation of a new technology forms a sort of bridgehead in the social system, and expansion takes place from this point outwards. The "integration" of the new technology... begins in a situation of at least partial acceptance. One can call up the image of colonialization: the triumphal march of the colonizers that conquers even the massive superiority of the "old cultures" can still be explained only by the fact that the latter are divided and ambivalent in the face of the invaders.[27]

Nonetheless in spite of all the successes (real ones as well as Potemkin villages) and in spite of all the promises, in public discussion the threat remains a persistent theme, and the dangers and risks are on public display. "Genetic engineering is eugenics!" shout the opponents; they argue that it will lead to worker selection and ostracize the handicapped. But because the process by which technology gains general acceptance is a "multifactorial process"[28] (i.e., cannot be derived from technology alone), surely formulas of this sort do *not* apply so simply. The pioneers of bioengineering also like to point this out, and they are doubtless right from the viewpoint of pure scientific theory. But in many cases they go one step further and say that the dangers—if any—lie elsewhere. Technology per se, in their view, is neutral. If it is abused, this is the fault of those who apply it. The blame lies with the users, human beings, society, politics, culture, stupidity, bad motives—only not with technology. In short *it's other people's fault.* Although the pioneers of bioengineering otherwise like to limit their view to the bound-

aries set by their own specialties, here they show an immense willingness to discover the power of cultural forces: "I would like to ask the
critics: isn't there, after all, a cultural heritage alongside the genetic
one? The overemphasis on the role of the gene seems to me excessive
and unjustified."[29]

And once again, according to two pioneers of reproductive technology,

Genetic manipulation cannot do harm all by itself. An entire state,
an entire country . . . managed, without manipulation of the gene pool,
to kill more than six million Jews. The extent of genetic manipulability
finally has to be seen in context. After all, the non-genetic manipulation taking place in all societies is a much greater cause for concern. We should also . . . note that in our society we talk constantly
about human dignity and at the same time we let news reports and
images from the media flood over us without reacting. The soccer
match in Brussels . . . happened not so very long ago. . . . Well-being
and disaster, happiness and unhappiness are caused . . . more by the
political, religious, and economic systems.[30]

To be sure the critics are not concerned with the role of the genes
but with the role of *genetic engineering* and its utilization, which is
quite a different thing in both logical and practical terms. But this
distinction is either not clear to the advocates or it is skillfully over-
looked. (It would be a delightful exercise for political scientists and
linguists to examine the patterns of argumentation, ambiguities, and
gaps in logic that can be filtered out of such passages. But no, let's not
pursue the topic any further here.) Let's limit ourselves to an analytic
description. In cases where many factors come together, the argument
over the "final cause," that is, the cause that turns out to be crucial, is
enormously explosive in political terms. "The starting and stopping points
of causal explanations can in many cases be justified only in pragmatic
terms, and can serve as the strategic variables of the social processes of
laying blame or absolving from blame."[31] The argumentative effect is
considerable. By attributing causes to others, ping-pong fashion, one
can make one's own role—for example, that of research and technology—disappear from sight altogether. One's role dissolves in misty
marginal areas. If people succeed in making technical dangers appear
to be part of human nature or culture, "they will be divested of their
social explosiveness. They will . . . be banned from the context of technical responsibility and responsiveness to outside influence."[32]

In case doubt and resistance still occur, yet another strategy of legitimation remains: *delegitimization*—of the opponent, naturally. One

can argue that the critics lack a rational basis, for example, by denying that they understand the facts of the situation, claiming they are motivated by their "thrills of horror,"[33] calling their fears irrational and finding them lacking in a sense of proportion, or finally—last but not least—imagining the irreparable dangers that a renunciation of technology would bring. After all, isn't it precisely the endless discussion of dangers that generates the worst dangers? This is the opinion of the president of the German Research Society, who suspects that "the greatest risks of biotechnology in the area of genetic engineering [lie] in the generation of horror scenarios."[34] The motto is always: Don't give in to hesitation and vacillation, don't give way to panic-mongering and fear! Instead face up to responsibility and turn your gaze bravely toward the future!

> Before the last reading of the law on genetic technology in the German Bundestag, the presidents of the scientific organizations turned to the caucus chairmen of the German Bundestag with a joint declaration on genetic technology. This declaration said, in part, "We regret that this discussion, in so far as it concerns genetic technology, is being conducted by many of its critics without the understanding of the facts that are necessary for assessment. . . . Fear and emotion are understandable reactions which policy-making and legislation must take into account. . . . But science must . . . try, on the basis of its understanding of the facts, to keep these emotions within proportions that are appropriate to the situation."[35]

Once again here are the words of two pioneers of reproductive engineering:

> Can we, and may we, make possible dangers an appropriate standard of measurement for the decisions we now have to reach in a responsible way? We are running the risk of becoming the victims of our own mistrust. . . . Fear has never a good counselor in coming to terms with new developments. We have to develop a resistance to panic; more composure is needed for the acceptance or rejection of options for our future. . . . In our fears we often fail to see the blessing of progress. . . . Why should we always only stare anxiously at the ugly side of the two-sided coin of progress?[36]

Linguistic Politics

The examples just cited lead us to suspect that linguistic politics is a broad field of action for creating agreement. Here a multitude of variants present themselves. I will select only two: the mottoes "accentuate the positive" and "ignore the negative."

As an introduction to the first category, we might take "Sunshine Genetics," the name of a market enterprise specializing in embryo transfer, or "Reproductive Freedom International," an agency that offers surrogate mothers to its customers.[37] Admittedly these are some of the cruder examples. More subtle is the concept of giving to a sperm bank, which immediately calls up an association with giving blood; in any case, isn't giving a good, humane thing, something that comes from the heart and shows altruism? The acronym for "artificial insemination donor," AID, functions in much the same way, as does GIFT, which stands for one variant of in vitro insemination. With so much helpfulness in evidence, the question might arise Why not donate a four-celled entity for science?[38] And then, in order to regulate the responsible use of this entity, an "embryo protection law" will be on the agenda.

This phrase may serve as the transition to the second variant, where disturbing elements can be maneuvered out of one's field of vision. For the embryo protection law in its current draft does not necessarily serve primarily to protect the embryo; rather it serves entirely different interests and goals, as formulated by a woman jurist during the hearing before the legal committee of the German Bundestag: "A law which permits the destruction of the embryo; which has the primary goal of obstructing its conception; and which incidentally aims to protect third-party interests or general dogmas, cannot be called an embryo protection law."[39] In the vernacular this kind of thing would probably be called deliberate mislabeling.

In other cases, concepts borrowed from the special terminology of science are used to package in respectable abstraction the core of what is meant in terms of behavior, thus making it practically unrecognizable. For example in vitro insemination is presented as an ideal method of the future to prevent the occurrence of severe inborn defects—which means, more precisely, the existence of handicapped *children*. In the area of prenatal diagnostics, "prevention" is a favorite word, used in phrases such as "prevention of Down's syndrome."[40] But the fact is that prevention is not possible in these cases, only an intervention (abortion) *after* Down's syndrome is detected. Also widespread are analogies of the following type: "Society ought to decide to eradicate muscular dystrophy (or) Tay-Sachs illness . . . in the same way that . . . measles has been eradicated."[41] This analogy elegantly sidesteps a few small differences. After all in the one case we are talking about inoculation for the purpose of protecting the patient's life and health and in the other about an abortion that definitively ends the life of the embryo.

Here it becomes obvious how the stretching of concepts can be used as an instrument for pushing objectionable areas out of our awareness when dealing with controversial projects. Because resistance arose to experiments that released genetic material into free circulation, the president of the German Research Association reminded us that "strictly speaking, even every birth of a human being [is] an experiment releasing genetic material into free circulation."[42] And an altered interpretation of the concepts of "predisposition," "prevention," and "therapy" can push back the borders of definition that have so far set limits on medical intervention.

The following critical comment was made by the biologist Rainer Hohlfeld: "This is shown, for example, in the transfer of genes for therapeutic purposes or the correction of immunological defects. . . . Almost all of the intended applications . . . amount to a lifelong prophylaxis of an inborn genetic disposition . . . , that is, to a project of positive eugenics which comes dressed up as the 'lifelong prevention of illness.'"[43]

Let us take as a final example surgical operations on the foetus, which constitute a medical growth area with a future. Because the foetus, as we know, does not exist alone, operations on it always affect the mother, too; they intrude on her body and her life and are not without their own risks. These, of course, frequently fade from sight in the relevant discussions, because the woman is put into the category of "foetal surroundings" and thus defined out of existence;[44] and in another connection she becomes similarly faceless, disappearing completely as a person, occurring only as a "therapeutic modality."[45]

Do we hear the objection that this is merely a choice of technical concepts, which may seem a bit odd to the outsider but is entirely normal in the medical profession? And that consequently this is not a strategy, it is neutral, nonpartisan, objective, scientific?

No, it is more. Whether it's normal or not, a strategy or not, language has consequences. Language directs our gaze, our expectations, and not the least our emotions (and it does so all the more as the world becomes more difficult to understand). Language is a compass, a definition of reality, a worldview, a set of instructions. If women are regarded as foetal surroundings, their rights can be ignored. If, when looking at severe birth defects, the embryos that happen to be attached to them are no longer even mentioned, then one can do with these embryos whatever seems necessary—to use the professional jargon once more, quality control with the mistakes sorted out.

The Price of Specialization

But stop! It is dangerous to assail the linguistic constructs of other disciplines, especially if one sits in the glass house of the social sciences—as though the social sciences did not have enough sins of their own: linguistic constructs that seem to have been freshly spit out by a data-processing computer, and comprehensive verbal campaigns that scale the highest levels of abstraction, with the last traces of color, point of view, and life mercilessly squeezed out of them. All those structures, processes, systems . . . but what has become of the subject, which has just been reclaimed from the other scientific disciplines?

This is meant to remind the reader that linguistic formulas like the ones that were presented from the field of medicine as examples do not function only as strategies of legitimation, deliberately invented for this purpose. It would be simple if it were only a matter of the behavior of a few researchers trying to camouflage their actions. The problems lie deeper. For behind them lies, as has been described above, a context of research characterized as a whole by specialization and abstraction, not only in the natural sciences but also in the social sciences. From the perspective of this specialization based on the division of labor, a woman, in fact, becomes "surroundings," a marginal condition—and the desire to camouflage something is not the only reason. The distinction between embryos and preembryos may be a political one made in order to create free space for embryo research—excuse me, preembryo research. But when one looks through the microscope, one sees in fact a four-celled entity or foetal tissue—not a person shouting loudly: "I am human life!"

This type of highly specialized research no longer has anything to do with people but with particles and elementary functions. Linguistic formulas like those cited above are thus not primarily the product of conspiratory interests, but the product of a type of thinking and acting inherent in the essence of the modern era, that is, in the rationalization process that brings, together with the Enlightenment, the dialectic of the Enlightenment. Medicine is only a particularly vivid example of this. In order to fulfill its most essential task, that of healing, and of course also to satisfy scientific curiosity—through these legitimate motives a development is taking place that is carving up the field into increasingly small plots. Only in the later stages does it become evident that in the process an increasing amount of abstraction and a deficiency of experience, especially sensory experience, are necessarily being created—together with all the dilemmas inherent in the logic of this behavior. The fact that this world with its lack of sensory experience then permits inscrutable word games and all kinds of strategies of legitimation is, as I've said, only part of the story. But this side effect is doubtless useful if one is bent on keeping one's vulnerable flanks protected and shielding new technologies from resistance.

4. Counterstrategies, or, On the Feasibility of Inherent Pressures

Pathways of normalization and constructions of acceptance open up the field for the process of gaining general acceptance of new technologies, breaking down resistance, and silencing doubts. However, the critics can also use these methods. It's a bit like the international espionage business. Once you've unmasked an enemy agent, you can turn him around and make him your own source of information. Once you've seen how the pathways of normalization and the constructions of acceptance are laid out, you can develop counterstrategies and perhaps also call for different blueprints.

To give an initial example: Once we see how the specialization of the modern natural sciences, based on the division of labor, lowers the thresholds of restraint and creates side effects one after the other, we must say, no, this specialization should not be abolished (because, as we know, it also has its advantages), yet specific counterbalances to it have to be created. To counter the pressure to specialize, we have to create pressure to think in terms that transcend specialties and anchor this pressure through incentives built into study, research, and practice, so as to direct people's gaze over and over again at the unwieldy, disturbing, uncomfortable whole, which is for these very reasons so necessary—at the reality that lies beyond the narrow boundaries of professional specialties. In other words we have to change the criteria and standards of measurement, selection procedures, and career ladders which punish those who look beyond boundaries and reward those who think only in terms of their own specialties. This must be done so that nobody will be able to say, ever again, what was said by a surgeon in his own defense during his trial for doing kidney transplants in connection with the sale of human organs: "I'm a technician. I take out kidneys, that's all. Questions of medical ethics have never interested me."[46]

Is this a utopia? It may be. But if it is not translated into action, if we don't find effective ways of requiring and encouraging people to look at the whole, in spite of every type of resistance—then the risks and drawbacks of modern medical technology will grow enormously in the foreseeable future. And this will not remain limited to the field of kidney transplants.

To make the utopia more concrete, let us take as an example the commissions and committees whose number increases apace with the discussion of the "risks and possibilities" of new technologies. Here the duty to think in terms transcending professional specialties can be anchored in a relatively simple way, through appropriate selection. If these commissions are to have any function besides that of legitimizing existing procedures, if they—presumptuous thought!—are to genuinely extend our knowledge of those questions to which nobody can yet provide the final answer, then it is necessary—if I may boldly say so—to staff these commissions in motley fashion. There should not be a predominance of insiders representing the interests of their profession, social class, or research area (and who could take this amiss?). Instead there should be people from various groups, with accordingly different experiences and types of knowledge. This does not necessarily mean randomly recruiting laypeople from the street. But even if there are no laypeople (and why not, actually—as representatives of the "responsible citizens"?), at least the majority should not consist of self-monitoring representatives of special interests. At the very least the circle of experts and counterexperts should be more variegated—that is, different majorities have to be created. Here is an example of what should *not* be happening:

> The Federal Medical Council has appointed a central ethics committee whose task is to ensure that basic ethical principles are adhered to in human-embryo research. Two physicians who are themselves working in the field of reproductive technology have written about the commission's make-up: "The central ethics committee includes practically only scientists who are receptive to human-embryo research."[47]

To make the issue even more concrete, let me share with you one of my own experiences in the working group on genetic research appointed by the federal minister for research. This group was to discuss not specialized questions from the natural sciences but—in the terms of their assignment—the "ethical or social consequences," the "significance for the self-concept of the human being," and for his "attitude . . . toward his fellow human beings."[48] But if this was the task, why was the choice of experts so narrow? Why was the distribution of guest experts initially so one-sided in terms of gender that it could indeed hardly be called a distribution (one woman, nineteen men) and drew public criticism in the media? (Subsequently the quota of women rose dramatically: two instead of one.) Why were only two social scientists invited, compared with nine natural scientists (some of them directly involved in genetic-engineering firms or research in molecular biology)? Most important why was there not one single delegate from groups representing handicapped people—although it is in just these groups that genome analysis and genetic therapy are being most heatedly discussed?

> An editorial in "Die Zeit" asked, "Why are only natural scientists, sociologists, theologians, and industrialists talking about these issues? Why not a couple of educators, a couple of poets, a couple of painters and musicians? Why not Robert Wilson, Jurek Becker, Walter Jens? Why not a couple of ordinary people? Why not—oh, it always occurs to us only at the last minute—why not make half of them women? On this, of all topics!"[49]

In the second place, if, given the circumstances of modern research, freedom of scientific inquiry easily becomes an instrument for shielding new technologies from critical questions asked by the general public—which at the same time opens the gate for more extensive utilization—then the postulate of freedom of scientific inquiry has to be reformulated. The point is to preserve freedom of thought *without* thereby providing a *carte blanche* for freedom of action.

> It (is) no longer possible to avoid the question of what the privilege of freedom of research is based on, if the conditions it used to depend on have been invalidated . . . the limits that are meant to curb applied science can not be drawn by applied science itself. They have to be imposed and enforced from outside. . . . The more its power increases, the more the policymakers will have to see to it that their voices do not sound like a faint echo of decisions that have been made elsewhere.[50]

In the third place, if the pioneers of bioengineering use certain strategies of legitimation in order to build up public acceptance and break down resistance, then it is necessary to look at these strategies of legitimation very closely and critically. To what extent are they based on truth and to what extent on fiction? For example, how are positive statistics produced, and what failures are possibly being defined out of existence? In the discussion of potential dangers, where are causes being shifted in a process of "causally throwing out the risks"?[51] Where are the gray areas, and what color-brighteners are being used? Which side effects remain completely invisible? For analyses of this sort, interdisciplinary dialogue is needed. At least equally necessary are the voices of the natural scientists who are familiar with reproductive and genetic engineering from the inside, with its promises as well as its dangers. They are the ones who can detect most keenly the falsifications in the reports, the gaps in the statistics, the touching-up of success rates.

Is this yet another utopia? Perhaps not entirely. The discussions being carried on already within the natural sciences among physicians, biologists, and geneticists indicate that a willingness exists. (Human genetics, as the following chapter will show, is a vivid example.) The more bioengineering—or, more precisely, some of its practitioners—pushes forward into broader areas, the more do their colleagues experience in their own professional practice what dilemmas and conflicts are being generated along with the promises. Thus it is of the utmost importance to give these experiences a forum—in professional journals, committees, congresses, and professional organizations. Structures for criticizing technology within the various professions must be promoted and consolidated through institutions. A touch of extraparliamentary opposition in the General Medical Council, reserving a few careers for members of the other camp, for the whistle-blowers, the protesters, the nonpioneers—what a motto that would be! Let's do something more modest. First it's important to ensure that the advocates of critical positions (male and female) do not meet with the gentle or not-so-gentle forms of repression, that they are not shaken off or put out of harm's way outside the institutions.

> Whatever has so far been able, with difficulty, to clear a path for itself by struggling against the dominance of the professions or business management has to be *consolidated by means of institutions*: counter-expertise, alternative professional practice, discussions within the professions and industries about the risks harbored by internal development, concealed scepticism. In this case Popper was indeed right: criticism means progress. Only where medicine opposes medicine, nuclear physics opposes nuclear physics, human genetics opposes human genetics, does it become possible for outsiders to discern and judge what kind of future exists here in the test tube. Making self-criticism possible in all of its forms is not a threat but probably the *only* way we can discover beforehand the error that might otherwise blow up the world around us at any moment. . . . Then it would

also be possible for technicians to report on their experiences in industry; at least they would no longer have to forget at the factory gate the risks they see and produce. . . . This institutionalization of self-criticism is very important, because in many areas if one lacks the requisite know-how it is impossible to recognize either the risks or the alternative means of avoiding them.[52]

Perhaps there is still a chance to offset the pressures inherent in the situation, by means of counterstrategies of this kind or other kinds. But if we are to do this, obviously we must recognize that they are not pressures in the sense of inexorable fate—and that, moreover, they are not inherent in things but originate in human actions. So our motto should be: "There are no longer any pressures inherent in the situation unless we let them, or make them, dominate." This does not mean that everything can be structured as we like. "But it does mean that the camouflage of 'pressures inherent in the situation' has to be set aside and, instead, interests, standpoints, and possibilities have to be carefully weighed."[53]

6 Brave New Health—Human Genetics in a Dilemma

1. The Health Trend

Jogging and granola, medical check-ups and fitness exercises for senior citizens, body building and partner massage, yoga for pregnant women, fasting cures, diets of every kind—the offerings are numerous, the fashions ever-changing, but the promise remains the same: health for everyone, for you and for me, health wholesale and nonstop. And not only health but, along with it, fitness and youth, well-being and achievement, beauty and strength, a long life with an optimal guarantee.

In the past it was religion that promised release from suffering; later on it was faith in one's people and the fatherland, race and the Reich, that required "toughening up the body of the people." This, too, has passed on. What remains is the person who exists here and now and his individual well-being. This is the goal toward which hopes and efforts are now directed. Now that faith in the hereafter is disappearing, health gains new significance, rises in value, and is transformed into the expectation of earthly salvation. "What can no longer be expected of the hereafter is now . . . being projected onto this life: freedom from cares and limitations, from illness and suffering—in the last analysis, bliss and immortality."[1] In a phrase salvation has been dethroned and healing has taken its place.[2]

Now in recent years it has been possible to explore new dimensions of this expectation of earthly salvation. Through progress in medicine, biology, and genetics we have succeeded in decoding the human being's genetic identity card, piece by piece. The more these technologies progress, the more precisely identifiable become the genetic aspects of health and illness. The human being's biological make-up can be investigated down to its elementary components. In the process, defects

79

in his make-up also become visible: what are called "anomalies" or "genetic defects," for example, inborn illnesses or predispositions to them. The list of such defects is long—from Down's syndrome to heart attacks, from diabetes to schizophrenia—and the envisioned goals are ambitious: what is still not decoded today will be decoded tomorrow.

Indisputably this opens up new possibilities for action at many levels. For example the information gained through genetic diagnosis can be used to protect oneself from genetic predispositions to illness, to offer specific support opportunities to children who have predispositions to certain illnesses, or to prepare the parents through special counseling for the particular challenges posed by a sick or handicapped child.

These are the possibilities. Whether they are actually used is admittedly another question. Moreover the experience of recent years has also taught us skepticism. In general wherever opportunities present themselves there are risks as well so we must pay attention to the other side of the coin—as far in advance as possible, not when it's too late. Thus the new promises force us to face new issues: What happens if the procedures of genetic diagnostics reveal more and more "defects" and "anomalies"? Can they help us become healthier? And what are the possible side effects of this health program?

2. The Situation of Groups at Risk

Let's first look at the situation of "groups at risk," that is, the men and women who are at increased risk of having a genetic illness. This includes, for example, people who have a close relative who has such an illness, or who already have a severely handicapped child, or who are at genetic risk because of certain life circumstances (for example, advanced age). To the extent that they are aware of this risk, such people face difficult questions regarding partnership and parenthood.

What kind of help can genetic diagnostics offer here? We might say it can offer "objectivized" assistance in decision making, that is, provide information on how great the risk is that a certain type of handicap will occur. (For example the diagnosis of Huntington's chorea, also known as St. Vitus's dance, means a 50 percent risk of illness for each child and 25 percent for each grandchild.) Or in the case of an existing pregnancy, the doctors can determine through genetic tests whether the embryo has a certain genetic anomaly (such as Down's syndrome, popularly known as mongolism).

This means that the doctors can make a diagnosis in many (but by no means all) cases. However in the realm of therapy, they have a great deal less to offer. (There is no medicine or operation to "heal" Down's syndrome, Huntington's chorea, or muscular dystrophy.) This is the dilemma of genetic diagnostics: the possibilities of diagnosis have moved far ahead of the possibilities of therapy.

For the clients of genetic counselors, this means that for some of them the findings doubtless bring relief—if they learn that the risk of the child's having a genetic illness is very low or if the results of prenatal diagnosis during the pregnancy shows that the embryo does not have a certain anomaly (such as Down's syndrome). But there is also the opposite situation, where the findings turn out to be unfavorable. In this case the clients receive not relief but a heavy burden, the burden of responsibility and the conflicts of decision making and conscience. Many clients in this group feel they are being left alone in their fear and despair. If they don't dare take upon themselves and their children the risk of a severe illness, then in many cases only two alternatives remain: either to forgo biological parenthood altogether or to hazard a "tentative pregnancy"[3] and counter the risk with prevention. What is meant by successful prevention can be illustrated by an example.

In an article entitled "The Prevention of Down's Syndrome (Mongolism)," human geneticist Werner Schmid writes, "In the year 1986 we dealt with 71 cases of Down's syndrome. In 52 cases we carried out the karyotype examination in . . . patients who were born alive; in these cases no prenatal tests had been made. During the same period 19 cases were diagnosed prenatally; the pregnancies were terminated. . . . Thus we knew of a total . . . of 71 cases. In 25 of them . . . the mother was older than 35. Of these 25 cases, 13 could be prenatally prevented. . . . Such prevention is quite successful in cases of pregnant women over 35."[4]

Thus genome analysis cannot help the affected parents to have a healthy child. And saying that it can help the affected parents to "prevent" the birth of handicapped children is a misleading euphemism. The basic problem lies in the purpose of the examination: "Prenatal diagnosis is usually carried out with the intention of identifying the handicapped child so as to be able to abort it 'in good time'—within the time period specified in § 218a of the Criminal Code."[5] One of Werner Schmid's colleagues, human geneticist Traute Schroeder-Kurth, has formulated the dilemma in plain words: "The illness is prevented by snuffing out the existence of the person who is ill."[6]

The Expansion of Groups at Risk

Now one may say that the groups at risk are a small and precisely delimited category, and so the problems occurring here are rare ones. But this is simply an error, for the extent of the groups at risk is not an objectively ascertained and static magnitude; it depends essentially on the current state of research in genetic diagnostics. In other words the more progress is made in decoding the human being's genetic map, the more anomalies become identifiable—and the more people then find out that they are carriers, in one way or another, of a potential "defect." For example in our mothers' and grandmothers' generations, there were also many women who became pregnant at age thirty-five or forty. But they would never have thought about whether the children growing within them might have Down's syndrome. Today, however, practically all women in this age group know about this risk—thanks to the mass media, and to the doctors' duty to inform their patients—and thus face new questions and conflict-ridden decisions.

Moreover research does not stop at those anomalies that involve severe physical or mental impairments. The inner logic of the procedures of genetic diagnostics requires that other defects in the genetic blueprint also become detectable—defects that do not affect the basic life functions but sometimes make the normal functions of everyday life uncomfortable or unpractical (e.g., extreme nearsightedness or slight metabolic disorders). These "merely uncomfortable" anomalies are now also, inevitably, drawing attention—and prompting, in their turn, questions and conflict-ridden decisions. Moreover predictive medicine deals not only with prenatal forecasts. A number of inherited illnesses do not become manifest until old age. People who know that such illnesses have occurred in their families are now being confronted with the possibility of having tests done that will reveal whether or not the illness will strike them, too, in future years. "If we also include the illnesses in which inherited and environmental factors jointly play a role, then predictive genetics becomes a problem for almost everyone,"[7] says Jörg Schmidtke, himself a human geneticist. What is becoming apparent here is a creeping expansion, or even inflation, of defects and recognizable anomalies, and the yardsticks and standards connected with them—an inflation resulting from the dynamics of the procedures of genetic diagnostics. Heart and circulatory diseases, cancer, allergies, endogenous depression, diabetes—genetic dispositions play a role everywhere. To put a fine point on it: *All of us are affected. All of us are risk carriers.*

3. Life as a Professional Patient

The expanding process of decoding the human being's genetic blue-print is making possible the early detection of an increasing number of human dispositions to illness. In the optimal case this can mean that appropriate medical intervention can prevent the outbreak of the illness or significantly moderate its course. But how frequent are these "lucky cases"? Perhaps they will be more numerous in the future—this is the promise. At present only one thing is certain: new conflicts and new levels of the problem are being generated.

New Levels of the Problem

As I've said before, for many symptoms only diagnosis has been possible so far because there is no effective therapy. Thus serious questions arise that develop tremendous dramatic force in the lives of the people who are affected. For example how do people live with the knowledge that in a few years or decades they will be struck by a severe illness, without any chance of escaping this destiny? How do parents live with the knowledge that their newborn child will inexorably be marked by such an illness? How do they live with the conflict of having either to tell the growing child about its genetic predisposition or deliberately remain silent about it for years?

Furthermore in many cases where the genetic predisposition to an illness becomes detectable at a very early stage, even before birth, the actual outbreak of the illness does not occur until middle age or old age. Until that time these people can lead perfectly normal lives. More precisely they can do so *if* they are not informed about their genetic predisposition. But in cases where this does happen, the chance to lead a normal life is subtly thrown off balance or even destroyed. The person's choice of profession, partnership, parenthood—all of his life decisions must now be made in the shadow of the looming blow of fate, which strikes the individual via mechanisms that are pitiless. Yet in other cases genetic diagnosis can only reveal a certain degree of probability that a certain illness will occur—but not whether it will *actually* break out. The question here is How do people live with such statements of probability which affect them existentially, in the literal meaning of the word? How can they deal with the resulting fear and insecurity? (Some, for example, will constantly monitor the state of their health in a search for signs of possible symptoms indicating the outbreak of the illness.) So here, too, the normal course of the person's life will be significantly disturbed—even for those who will not in fact become ill. For this group of people, it is definitely not the illness but the *information* provided by genetic diagnostics about a possible illness that inflicts lasting damage on the normality of their lives.

And then there are those who know, because of their family history, that they may have an inherited illness. This is certainly not easy to live with; it leaves its mark on all of one's life plans. But what if a new test suddenly appears on the market that can provide a clear result— yes, what then? Whatever the result of the test, it touches the core of the person's life plans. A person who is affected has said,

> The result of undergoing a genetic test will have radical effects on the psychic and psycho-social make-up of the person seeking advice. And as soon as the person at risk learns the result, these consequences will become an indelible component of his code for living. Such momentous genetic information can be neither ignored nor effectively suppressed. . . . Thus in reality it is not an abstract piece of statistical data that changes; one's own personality experiences a change, one sees oneself differently and is seen differently by others. . . . For a person at risk . . . the complicated psychic balance whereby he tries to negotiate the tightwire between hope and fear is now shaken because a test result has once again overturned the foundations of the life plans he has had till that time—namely, his self-concept as the carrier of two potential destinies—and makes possible a clear categorization (positive/negative—healthy/ill), or forces it upon him.[8]

Now in the case of some illnesses, the probability or severity of their occurrence can be influenced by an appropriate life regimen (diet, not smoking, avoiding alcohol, avoiding stress, etc.). This means that the person who learns of his genetic risk factors, his special susceptibilities—for example, a predisposition to diabetes or heart attacks—can adjust his life accordingly and thus perhaps avert destiny. This goal is surely reasonable and generally accepted. But the situation might become less reasonable if we look more closely at the ratio between means and ends. Thus probably most people would readily accept a small alteration in the way they live (for example, a diet that is easy to follow) if it meant being able to prevent the outbreak of a severe illness. But what if very extensive restrictions on one's way of life are required in order to prevent an illness that will not occur for several decades and even then perhaps only with a certain degree of probability? What is greater in this case, the gain or the loss? How can one calculate the quality of life and the quantity of life in a single equation?

Here the dialectic of the Enlightenment becomes concrete: the model of the individual's responsibility for himself can give rise to the self-imposed compulsion to live one's life according to the dictates of genetic information. The autonomy promised by genetic knowledge ("Take destiny into your own hands! Prevent illness!") imperceptibly turns into a dependency on the experts, which produces imperatives for increasingly detailed types of action. Knowledge of one's genetic make-up affects one's self-image in a fundamental way and makes deep inroads into people's life structures, life planning, and the processes of their everyday lives. It subjects them to the restrictions imposed by medical care, concepts of prevention, and monitoring procedures.

Diagnosis Without Therapy: Huntington's Chorea as an Example

So far I've only sketched in, very abstractly, various types of problematic situations resulting from the use of genetic engineering. I've barely touched upon the whirlpool of fear, helplessness, and despair, the insoluble conflicts of decision making that people who are affected can get caught up in. To illustrate more graphically the dynamics of these conflicts, I would like to relate a fictional case history.[9]

I have chosen as an example Huntington's chorea, a degenerative brain disease that generally occurs in the fourth or fifth decade of life and inevitably leads to premature death. No therapy has been found for it so far. As a result of family studies, it has been known for some time that this is an inherited disease with a 50 percent probability that the child of a carrier will have the disease.

For several years now it has been possible to ascertain by diagnostic means, with a high degree of certainty, whether a given person having relatives who are carriers will or will not come down with this pernicious illness. Because the location of the "Huntington gene" has now been narrowed down, it ought to be only a question of time before it can be clearly identified. This, in turn, will not only make possible a reliable test for carriers but will also open up the possibility of investigating in detail the function of this gene and developing a therapy on the basis of this knowledge about its biochemical effects.

But until such a therapy exists and until the testing procedure is perfected, the relatives of people having Huntington's chorea will be thrust into wrenching conflicts. So far they know only that there is a 50 percent probability (if one parent already has the illness) or a 25 percent probability (if only one grandparent has the illness) of being carriers of the Huntington gene. Most people in this situation create for themselves a precarious balance between fear and hope, in which hope finally wins out. This affects not only their own life plans but also the question of whether they ought to have children—who would then, if the parents themselves should turn out to be carriers, have a 50 percent probability of inheriting the Huntington gene. Genome analysis disrupts this precarious balance.

Let's take a look at the newly arising conflicts in the life of a fictional person, Birgit Binder. Ms. Binder is thirty-five years old and has been married for ten years. She and her husband have decided not to have children because her father developed the illness shortly before their marriage; he died recently after spending several years in a psychiatric clinic. But in spite of contraceptive methods, Ms. Binder has become pregnant and is now in her second month. Because of her fear of a "final" diagnosis, she has not yet had the "Huntington test" done. There is still time to have an abortion. What should she do?

Should she have amniocentesis done (the risk of its causing a miscarriage is about 1 percent) to find out whether her child is a Huntington carrier? A positive diagnosis would make it easier for her to decide to have an abortion. But at the same time she would also know that she herself will die of Huntington's chorea in years to come, if an effective therapy is not developed very soon—which is very improbable. If the diagnosis is negative, she could carry the child to full term fairly free of anxiety (the probability of its being a Huntington carrier would then be 5 percent, because the test procedure has not yet been perfected). Her risk of being a Huntington carrier would still be 50 percent . Nonetheless the news that her child was not affected would probably encourage her to immediately have her own genetic make-up investigated, too. A negative diagnosis would free her from her long years of anxiety. A positive diagnosis would—apart from its shattering effect on herself—thrust her anew into the conflict of whether to have an abortion after all, because there would be a high probability of her becoming ill while the child was still young.

Probably Birgit Binder is now also thinking about her father, with whom she had a close relationship. She knew him as a cheerful man with a zest for life, who led an entirely normal life until the age of forty-five. Was this life not worth living? What would in fact justify having an abortion, even if it turns out that her child has inherited Huntington's chorea, even if there is no therapy in the future? And anyway—she suddenly thinks—where is the borderline? Many people die between the ages of forty-five and fifty-five of heart attacks. In a few years there will probably be tests to reveal the genetic risk factors for having a heart attack at an early age. Should we abort embryos whose chance of dying of a heart attack before the age of forty-five or fifty or fifty-five is 50 percent or 70 percent or 90 percent?

Of course Birgit Binder is now also thinking about recent developments in medicine and genetics. Perhaps she now reads one of the German industry's large-scale advertisements for the "great opportunities of genetic technology," which promise that "the treatment of incurable illnesses is moving into reach."[10] Should she now set her hopes on this? Now that the scientists are, so to speak, holding the primeval substance of heredity in their hands, they will gain new insights into the processes of illness within the foreseeable future. This could also result in new concepts of therapy. What kind of conflict would she have to live with if she had an abortion today, but in a few years its illness could perhaps be cured?

This is genetic engineering's version of Russian roulette. It is a game of chance, and at stake are life, health, and death—no, more, the *decisions* we are being asked to make now. It is a game that has many unknowns, yet has to be played according to the rules of an either/or logic, for in the end only clear decisions are permitted: yes or no to a test, yes or no to an abortion.

Is Knowing Better than Not Knowing?

"Finding out genetic data causes a crucial change: the state of not knowing is over for good, and knowing is a challenge to act responsibly."[11] In this sentence from an article on genetic counseling, an important premise becomes visible: genetic diagnostics presumes a human being who can weigh his alternatives rationally and cope with highly complicated medical findings and the ethical problems to which they lead. But it's uncertain whether the questions that arise here are even susceptible to a rational approach, or how average citizens are supposed to deal with the deluge of information flooding in upon them—that is, the medical lay public, that already feels overtaxed and helpless in the face of the experts' public controversies over genetic engineering. "Genetic counseling assumes that the advice-seeker is able to find an appropriate attitude and type of behavior in his specific situation."[12] How often is this improved orientation actually achieved? How often are new conflict situations generated instead, which throw the affected person's life plans off-balance? To rephrase the question, is it correct to make the assumption inherent in genetic diagnostics' paradigm of action—that knowing is always better than not knowing?

No longer are all human geneticists prepared to answer this question with an automatic yes. Caution and skepticism are in evidence when, for example, Werner Fuhrmann writes,

> Twenty or thirty years ago we developed genetic counseling based on the perhaps rather naive assumption that more knowledge and better knowledge is always good, and that we only have to make the most precise diagnoses possible, calculate the most precise prognoses possible, and make them understandable to the people who seek our advice, in order to help them make a decision which is right for them. Today we know much more clearly that this is certainly not always the case.[13]

The dilemma is much more starkly formulated by Jörg Schmidtke in an article with the telling title "The Loneliness of Facing the Truth."[14] Schmidtke begins by pointing out the burdens inherent in knowledge and the freedom of not knowing: "The scientist easily loses his sense of the significance of this freedom. After all, his job is to create knowledge. For the consumer of knowledge, however, this knowledge can quickly become a burden, especially if it affects him personally. For, unlike objects which can be given away or thrown away, unpleasant knowledge, once gained, makes trouble—at best, it can only be repressed." Schmidtke lists the many questions that come with a reliable prediction: "Does it always make sense to know about personal risks? Can such knowledge be used to develop more self-aware ways of living? Is a person happier knowing a thing, compared to supposing or hoping it? We have to assume that in years to come we will be able to know more and more about our individual futures. But how will we be able to live with the opportunities and fears implicit in this knowledge?" In conclusion he quotes from Friedrich Schiller's poem "Cassandra":

> What purpose does it serve to lift the veil
> When destiny looms nearby?
> Only error is life,
> And knowledge is death.

4. Voluntariness or Pressure

Now it is pointed out over and over that the use of the methods of genetic diagnostics should be based strictly on the principle of voluntariness: nobody should be forced to undergo such tests. The question is only How does this voluntariness look in practice? Can it be maintained in the long run, or will it come increasingly under pressure from different directions?

The example of prenatal genetic diagnosis may, once again, show us how matters could develop. Here, too, nobody is forced to expose himself to this information; in other words no woman is forced to undergo prenatal diagnosis. But the doctors are duty-bound to point out the option of prenatal diagnosis to patients at risk, and this category includes, for example, all women older than thirty-four. An informational process of this type, however, always includes a certain "element of demand."[15] The loss of one's ignorance is not neutral; it is a social fact of a peculiar kind. For scientific and technical options in themselves irreversibly change the field of action. Unless they entirely forgo medical assistance, women in fact have information continually forced upon them. This "forced information" is the drawback of the demands for personal freedom of decision. In an individual case this freedom may also include not learning about options for action and their results. But even a deliberate avoidance of, for example, prenatal diagnosis confronts the pregnant woman with the possibility of such a diagnosis, the potential risks for the foetus, and the consequences of this avoidance.

And what is true of prenatal diagnosis also applies to the other areas where the methods of genetic diagnostics are applied, such as, the area of predictive medicine. The basic fact is always that *the information per se has about it an element of social compulsion.* "In a certain sense it is impossible not to take new possibilities into account. One may not be interested, but the conditions of life are changed anyway. Actions and responsibilities are reorganized in society in the light of new insights and new technical options. The possibilities of identifying genetic risks, predicting future illnesses, or selecting out defective foetuses before birth, are irreversible—and they have to be taken into account whether we want to or not."[16] Once again doctors' duty to inform their patients plays a crucial role here. A consequence of this duty is that nobody who goes to the doctor has the opportunity to close her/his eyes to the new possibilities that are developing.

Preventive Pressure, Preventive Mentality

From these examples it is becoming evident that social pressure can arise at many levels, a pressure that does not need the measures of state repression yet in the end generates a pressure to be healthy. One particular reason for this is that in our society health is a high-ranking value which, so to speak, carries its own legitimacy. "Health fuels voluntariness and makes it comply with 'necessity.'"[17] Thus the talk about voluntariness misjudges and abbreviates the relationship between social values and new technologies to a purely private relationship; it does not acknowledge the new forms of social pressure that are arising here. "Today, pressures tend to become entrenched in a more subtle and indirect way. They loom in areas where the balance between the affected persons' interest in freedom and the public interest in the control of risks has not yet been decided through a clear assignment of values; here they can creep into society under cover of the general preventive mentality and through the pressure of their obvious rationality."[18]

This preventive mentality is already visible today in the new way of dealing with pregnancy which is becoming entrenched as the development of medical technology offers more and more possibilities for intervention. Today we can examine the foetus, monitor its condition in detail, identify defects and dangers at an early stage, even operate on it while it is still within the mother's body. This development leads us into difficult, if not insoluble, ethical questions and problems of assessment. The more visible and treatable the growing foetus becomes, the more massive limitations are imposed on the women's self-determination. "The extension of prenatal diagnosis and the consequence inherent in it, the duty of treating the foetus, is annulling, step by step, a woman's right to refuse the routine treatments offered by the doctors with their sophisticated technological equipment."[19] What is being opened up is a "field of preventive pressures that in effect abolish a woman's self-determination regarding her body and way of living."[20]

Preventive Medicine as a Precept of the Individualized Society

This preventive mentality has a good chance of winning general acceptance particularly because it conforms perfectly with the biographic models that modern society demands and encourages. One of its essential features is a tendency to individualization. This means "that the individual's biography is cut loose from prescribed definitions, is open and dependent on decisions, and becomes a task to be fulfilled by individual actions. The proportion of possibilities for living that are in principle closed to decision is shrinking, and the proportion of the self-creating biography which is open to decision is growing." In the individualized society, the individual must learn "to view himself as a center of activity, a planning bureau for the course of his own life, his capacities, orientations, partnerships etc."[21]

From this perspective the ego ideally becomes the midpoint of a complicated system of coordinates that encompasses many dimensions—from education and the job market to health insurance and pension plans—and must be constantly updated and revised. So far the demands of the job market have been a central axis of personal life planning. The offerings of genetic diagnostics conform to this tendency to rationalize the way one lives and to extend it. The possibility now arises of including genetic information (for example a predisposition to heart attacks or diabetes) in one's personal life planning as points of connection and guideline data. Preventive methods of self-protection are an "aspect of 'self-management' which is expected of the modern individualized human being. If a methodical approach to living becomes generally accepted, from planning one's educational career . . . to planning 'successful ageing,' then safeguarding health through preventive measures will gain high relevance."[22]

The modern person takes his fate into his own hands. He plans, he looks ahead, he monitors and optimizes. He no longer obeys God or the stars, and now his genes tell him how to organize his life. Two doctors in the field of reproductive medicine, writing about genetic engineering, have expressed this as follows: "A person's own knowledge about the genome should make him shape his life responsibly." Gene charting should be advocated, they continue, "if genome analysis should reveal that large numbers of people are carriers of inherited predispositions that are more vulnerable to environmental pollution. The public health system will have to show . . . a stronger interest in these people—even if the price is that carriers of these inherited predispositions will have to be informed about their higher susceptibility."[23]

Here the authors are no longer reflecting on whether and how environmental pollution can be reduced. The responsibility is placed squarely on the individual: it's his hard luck if his genes turn out to be vulnerable to pollutants. If this happens, he/she must simply act responsibly. They're *his* genes, so it's *his* responsibility. This is the logic of the individualized health program.

How Nondirective Is Nondirective Counseling?

In order to learn from history, in order to prevent a repetition of eugenics, human genetics has committed itself to the model of nondirective counseling. This means that the aim is to inform people (about risks, causes of illness, and possible therapies) but not to give advice. The client is supposed to make his/her own decision, with the help of the knowledge offered by genetic counseling.

So much for the norm. But, as is well known, norms don't always correspond to reality. So the question is How does it look in practice?

Many pregnant women don't even go to a genetic-counseling center in the first place; instead they are informed by their gynecologists about the possibility of genetic diagnostics. But in many cases this information is only fragmentary, and possible emotional consequences (e.g., effects on the pregnancy or the consequences of a late abortion) are often left out altogether. In many gynecologists' offices there is neither time nor space for women's doubts and fears. "Often in counseling sessions people are told how to act rather than being helped to come to a decision."[24]

> In an interview on this theme, a woman recalled, "He wrote a refer-
> ral for me, and that was it." Another one said, "After I asked a few
> more detailed questions, he said, 'If I were your age, I'd have it
> done—just to be on the safe side.'" Another gynecologist said to a
> 35-year-old patient, "A woman of your age—absolutely! From the
> age of 35 onward, you *have* to."[25]

In many cases, if a pregnant woman decides not to undergo genetic diagnosis, her gynecologist gives her a form to sign. On it she must certify that she has been fully informed about the possibilities of testing and has decided, in full awareness, not to be tested.[26] This procedure protects the physician from possible lawsuits by patients. But at the same time it puts the responsibility—which can easily become blame—on the patients if the child should be handicapped. The combination of new technologies and the legal regulations they generate creates a situation that mobilizes fears, which in turn augments the patient's willingness to undergo the examination. If women nonetheless hesitate, many staff members in the clinics react with incomprehension or blunt demands ("What's there to think over? After all, you were born in '53.").[27] And what happens if the woman decides to have the test done and a genetic anomaly is discovered? A book about pregnancy counseling and perinatology states, "A chorion villus biopsy or amniocentesis made on the basis of a genetic indication should be carried out only if the parents are willing, should the examination finding be pathological, to draw the therapeutic consequences (interruptio)."[28] Thus are dilemmas created: first fear is mobilized; then to calm the fear, there is a test—which the woman, however, should undergo only if she's going to have an abortion if need be. Each step, taken in itself, is neutral, but taken as a sequence they amount to a clear directive for action. And sometimes this directive is formulated explicitly. As human geneticist Christa Fonatsch says in an interview, "Of course there are borderline situations in prenatal diagnostics. Not all genetic defects . . . are so severe that an abortion would have to be recommended."[29] *Not all*—does this mean that in other cases it *does* have to be recommended? How nondirective is such a recommendation, which is also made publicly as it circulates through the media?

In this connection we must also look at the history of human genetics. And by history I am not referring to the era of eugenics and "racial hygiene" that ended in "selection and eradication," mass murder and extermination. I am referring to a more recent period. In 1972, when human-genetics counseling centers were introduced into the FRG, the following reason was given: "The most important task of the genetic counseling centers is to reduce the births of handicapped children within the limits of our possibilities."[30] Now, human geneticists do point out that there has been a nearly total transformation of goals within human genetics in recent years.[31] This may indeed be true. But whether the effects of earlier goals on public consciousness, the cost-benefit calculations, and the expectations connected with them can thus be obliterated—that is another question.

Thus even if genetic counseling in the counseling centers in fact remains nondirective, the framework of action in which genetic counseling exists as a whole, the preconditions and surroundings within which its offerings are presented and perceived—all of this is at many points *not* nondirective, *not* neutral. On the contrary inherent in it are many directives that encourage certain decisions and relegate others to the margins.

In the end only one question remains: How nondirective is genetic counseling itself—that is, not what gynecologists or staff members in the clinics say or what human geneticists might state in newspaper interviews but rather what actually happens in a specific counseling session in the counseling institutions that have been specially set up for this purpose? Experience shows that in many cases it is difficult to remain wholly neutral, for the simple reason that the clients *want* advice or a recommendation; they feel helpless and probably expect to be relieved of some of the moral and psychic burden of difficult and conflict-laden decisions ("The counselor agreed it would be better this way."). Besides it's by no means easy for the counselor to keep himself and his own feelings entirely apart from the situation, especially if there seems to be a very clear idea of what would be best for all concerned. As a woman counselor said in an interview, "It's often hard for counselors to be value-free. Oh, I know I ought to be value-free. But if you see a mother who's living on welfare and pregnant with her third child, and the man's giving her no support, and the child has sickle-cell anemia on top of that, it's pretty hard not to steer her toward an abortion. Then I think to myself, she's got enough problems."[32] These words make it obvious that every counseling session amounts to an incursion into the client's life history and that the counselor's personality cannot remain uninvolved. To put a fine point on it, "The concept of an 'ethically neutral' counselor in medicine is a fossil from the positivist age of naive ideals concerning science."[33]

5. Changing Images of Man and the World

The unfolding analysis of the human genome is increasingly investing man with the role of the creator of his own nature. His biological make-up is being opened to decision; it is becoming subject to planning, alteration, and correction. Thus the question of the blueprint inevitably arises: What can be left as it is? What needs to be corrected? Which defects are tolerable and which ones aren't? What should be improved and how?

Such questions, which are inherent in the logic of genetic engineering, make human life the raw material for directive incursions of various kinds (correction, therapy, foresighted planning, and "avoidance" through abortion if necessary). Such incursions can doubtless remove or at least diminish suffering. But in many small steps, each of which seems plausible in itself, they also clear the pathway for an instrumentalistic rationality that in itself knows no limits. A new attitude toward human life is in the offing: the value of life is becoming relative in terms of its genetic substratum alone (for example, in prenatal diagnostics or especially in preimplantation diagnostics). In the extreme case life becomes disposable material (embryo research which uses up the embryos; embryo splitting) or a storehouse of spare parts: children are being conceived in order to provide the genetically required substance for a family member who is ill—and they can be disposed of via abortion if they turn out to be "useless" for this purpose.

By way of illustration, take the following sketch of a possible case. A family has a beloved ten-year-old child who is suffering from a se-

vere form of aplastic anemia (Fanconi's anemia) and is already at a
lethal stage of this disease because her bone marrow can no longer
produce blood cells. One sister is healthy but cannot become a bone-
marrow donor because her cells would not be accepted by her sick
sister's organism. The mother decides to become pregnant again, and
now consults the experts:

In the first place, the child should not have Fanconi's anemia. (There
is a 25% risk of this predisposition recurring. A fractional analysis
of the chromosomes in the child's tissues can confirm or rule out this
illness.) In the second place, the child should be a boy, because the
mother already has two girls; in the chromosome analysis, the chromosomes indicating gender would be recognizable in any case. In the
third place, this boy should be capable of being a bone-marrow donor. This means that his bone-marrow cells would not be rejected by
the organism of his sick sister.

The mother is determined to have an abortion if these requirements cannot be met. For this reason, a chorion biopsy is done. The
chromosome examination shows that it will be a healthy boy who
does not have anemia but cannot be a bone-marrow donor. The family
accepts the child because it is a boy. The woman becomes pregnant
once again, and this time it is a healthy girl—who also, however,
cannot be a bone-marrow donor. She is aborted. The woman becomes pregnant again. . . . The story can be continued *ad lib*.[34]

In the meantime this mentality of do-it-yourself genetics is clearing the
way for the stigmatization and exclusion of people whose "genetic material"
includes obvious defects. Now that chromosome anomalies are drawing
increasing attention and handicaps are being increasingly defined as
slips or even mistakes of nature, this cannot help but affect the situation of the handicapped. First of all we can expect that the general
attitude toward handicapped people will change imperceptibly: toler-
ance and acceptance will decrease, and categories of "wrongful life"
will take root (in public consciousness, the legal system, the insurance
business). Moreover there probably will also be a change in the social
definition of "handicapped" or its social label. Gradually people are
starting to regard being handicapped as not a burden imposed by fate
but as *an event that can and should be avoided*. Thus the possibilities
for action being opened up by genetic engineering are turning into
obligations to take action, not through legal ordinances but through
the creeping erosion and change of value judgments, attitudes, and
expectations (at least in the Federal Republic of Germany). A line of
development might arise that looks like this: parents expecting a child

experience growing pressure from those around them to be sure to take advantage of the offerings of prenatal diagnostics; if they don't, they are openly or covertly written off as irrational, suspect, or even irresponsible. Especially hard-hit the parents who deliberately decide to have the child even though it is handicapped; those around them will grant them less understanding and less support than they urgently need. And this also affects parents who are not members of groups at risk and therefore don't have the relevant tests made and then have a handicapped child anyway or parents of children whose handicap is not genetic but originated during or after birth. Whatever the actual causes of the handicap, the social situation is determined by a chain of associations that seems to be forming in the public consciousness: handicap—genetically determined—therefore avoidable—therefore "it's their own fault."

On the Road to Biological Reductionism?

Through the sum total of such changes, a new image of man and the world may arise that defines the human being entirely in terms of his biologic predispositions. An example of this is the following sentence used by the geneticist Müller-Hill to describe the future of genetic diagnostics: "The variety of human beings, which reveals itself at present only in their behavior, can then . . . reveal itself in their DNA. The 'being' of the philosophers is changing into DNA."[35] This amounts to a biologic reductionism that deduces everything about a human being— all of his abilities and disabilities, all of his sufferings and achievements, all of his ways of thinking and behaving—from his genetic predispositions. Inherent in this is, specifically, a narrowing of our pluralistic conception of illness that entirely ignores the social and environmental circumstances. For example alcoholism, rape, and compulsive gambling are already being attributed to genetic dispositions as though one no longer had to ask, in such cases, about early childhood experiences, the crises the individual has lived through, and current triggering mechanisms such as unemployment or divorce.

Surely such biologic reductionism is not inherent in genome analysis. But it is also not far from it, as is shown by statements made by prominent geneticists. And even if it turns out to be untenable in terms of scientific theory, it nonetheless has consequences. One of the most famous sentences in sociology is the so-called Thomas Theorem: "What men define as real is real in its consequences." Wherever biological reductionism occurs, it does not remain mere theory; it also implies a concept of practice. It includes clear directives for political action. For example, if one assumes the determinative power of the gene, then the demand for equal opportunities in the educational system becomes less generally acceptable. Whereas in the past social reforms seemed necessary, we can now place our bets on the improvement of genetic predispositions. In short the danger of biological reductionism lies in the fact that instead of seeking social solutions people will start to seek only technical solutions. An example of this type of thinking is the following passage from the geneticist Julian Huxley: "If we want to make any great progress in terms of national and international productivity, it is surely not enough to tinker about with social and political symptoms and patch up the machinery of world politics . . . or improve the educational system; instead, we must increasingly change our goal to lifting the genetic level of man's mental and practical capacities."[36]

The Utopia of a Society Free of Suffering

The physicians working in the field of human genetics frequently encounter people who have to bear enormous burdens. Some people are able to accept these burdens and grow through the experience,[37] but others are destroyed by them. Now, because the doctor's task is to remove or at least diminish suffering, it is entirely understandable that doctors try to help avoid genetically conditioned suffering. Admittedly in the course of genetic engineering this leads us to the undertow just described, where the main aim is to continue expanding and using the diagnostic instruments of genetics. The result of this development is that a technically truncated worldview is becoming increasingly accepted by the general public, one that no longer views the human being in his entirety as body and soul but instead as the mere carrier of potential "defects." This worldview defines illness as negative per se, a destructive and frightening thing that must be avoided in every case.

Other cultures knew another way of dealing with illness and suffering (and some of them still do today).[38] Illnesses, as Novalis wrote about two hundred years ago, "are years that teach the art of living and form the character." This Romantic poet was by no means bursting with health—nor were Beethoven, Hölderlin, Kafka, Schiller, and so on. Novalis, who died of tuberculosis at the age of twenty-nine, knew well that "pain can turn a man to stone," but he also knew that "the man who flees pain no longer wants to love."

The question is not whether handicaps are bound up with pain and suffering. That is undeniable. Instead the question is whether a technological approach to handicaps is the right one and where the continual expansion of the procedures of genetic diagnostics is leading us. At this point we might recall some sentences of the Peruvian poet Vargas Llosa. The ideas he formulated about politics may be no less applicable to modern medicine with its high-tech trappings: "The fact that one can cause so much damage with the best intentions in the world, and at the price of a boundless willingness to make sacrifices, is a lesson I'm continually aware of. . . . It has taught me how narrow is the dividing line that separates good and evil, and what care is necessary in order to . . . solve social problems if we want to avoid having the cure cause more damage than the illness."[39] Genetic engineering has no ready solutions for the dilemmas it generates. These dilemmas are passed on to politics and society, especially to the many individual men and women who have to try to cope with them in their private life histories—which they may or may not manage to do. If the problems that arise here turn out not to be subject to limitation, if they cannot be mastered, then this would be—in the words of the former president of the Federal Constitutional Court, Ernst Benda—"the destruction of human dignity under the banner of humanity."[40]

Notes

1 An Interview by Way of Introduction

1. The method of realistic constructivism cannot be found in any social-science textbook. It is an original method that I developed especially for the purposes of this introduction. Not wanting to present it to this world unclothed or to violate the proprieties of science (perhaps even hoping to start new schools of thought?), I tried immediately to find an appropriate label.

   I call this method *constructivism* because it is not based on any interview that ever actually occurred in this form—as a question-and-answer game played by two people in a certain place at a certain time. But I call it *realistic* because it is not a game of fantasy: it deliberately takes up and strings together repeated questions that many people in many places and on many occasions have asked me—at panel discussions and commission meetings, in seminars and private conversations—questions by colleagues and students, politicians, friends and acquaintances, and natural and social scientists. Thus seen this interview is not a work of fiction but a concentrated rendering of reality.

2 The Technologies as Seen by the Social Sciences

1. Wolfgang van den Daele, *Mensch nach Maß* (Munich, 1985a), 11.
2. Neil Postman, *Die Verweigerung der Hörigkeit* (Frankfurt, 1988), 11.
3. Regine Kollek et al., *Die ungeklärten Gefahrenpotentiale der Gentechnologie* (Munich, 1986), inside jacket.
4. Wolfgang van den Daele, "Politische Steuerung . . . ," *Forum Wissenschaft*, 1 (1987): 42.
5. From an interview with Norbert Elias, quoted in *Süddeutsche Zeitung*, 3 August 1990.
6. Ulrich Beck and Wolfgang Bonss, "Verwissenschaftlichung ohne Aufklärung?" in *Weder Sozialtechnologie nach Aufklärung? . . .* , eds. Ulrich Beck and Wolfgang Bonss (Frankfurt, 1989), 24.
7. In vocational research this is called the "double-purpose structure" of professional work. See Ulrich Beck, Michael Brater, and Hansjürgen Daheim, *Soziologie der Arbeit und der Berufe. . . .* (Reinbek, 1980), chap. 8; specifically on the aspirations of the genetic researchers, and see Mathias Greffrath, "Der lange Arm von Chromsom No. 7 . . . ," *Trans Atlantik* 12 (December 1990): 14–28.
8. See, for example, Ortrun Jürgensen, "Gedanken zur manipulierten Fruchtbarkeit," *Wege zum Menschen* 2 (March 1990): 56–62.
9. Unfortunately such cases cannot be found in the medical textbooks. I have

the following case history thanks to a story told to me by Traute Schroeder-Kurth: a married couple came to a genetic counselor for prenatal diagnosis. Because they already had a daughter with an albino anomaly, the prenatal diagnosis was supposed to show whether there was a risk of this anomaly in the wife's current pregnancy as well. In the course of further counseling, it came out that the husband already had a new partner and wanted to separate from his wife. He wanted prenatal diagnosis so as to have an "objective" argument, in case a genetic defect was detected, which he could use to persuade his wife to have an abortion.

10. Traute Schroeder-Kurth, "Vorgeburtliche Diagnostik," in *Das manipulierte Schicksal*, eds. Traute Schroeder-Kurth and Stephan Wehowsky (Frankfurt, 1988b), p. 42.
11. Barbara Mettler-Meibom, "Mit High-Tech zurück in eine authoritäre politische Kultur?" in *Essener Hochchulblätter* (Essen, 1990), p. 61.

3   From the Pill to the Test-Tube Baby

1. Substantial sections of this chapter are an expanded and updated version of the essay "Von der Pille zum Retortenbaby: Neue Handlungsmöglichkeiten, neue Handlungszwänge im Bereich des generativen Verhaltens" (From the Pill to the Test-Tube Baby: New Options, New Pressures in Reproductive Behavior), published in Kurt Lüscher, et al., eds. *Die "postmoderne" Familie* (Constance,   1988), 201–15.
2. See, e.g., Hans Harald Bräutigam and Liselotte Mettler, *Die programmierte Vererbung* (Hamburg, 1985),   51, 66, 105, and 129.
3. Ibid., 112.
4. Ibid., 144.
5. Wolfgang van den Daele, "Technische Dynamik und gesellschaftliche Moral," *Soziale Welt* 2/3 (1986): 151.
6. Hans Jonas, *Technik, Medezin und Ethik* (Frankfurt, 1985), 11.
7. Ulrich Beck, *Risikogesellschaft* (Frankfurt, 1986), 335 f.; underlined in the original.
8. Standing Deputation of the German Jurists' Congress, 1987.
9. Peter Weingart et al., *Rasse, Blut und Gene* (Frankfurt, 1988), 16.
10. Laura Nsiah-Jefferson and Elaine J. Hall, "Reproductive Technology," in *Healing Technology: Feminist Perspectives*, ed. Kathryn Strother Ratcliff (Ann Arbor: 1989), 93–117.
11. Ute Frevert, "Fürsorgliche Belagerung," *Geschichte und Gesellschaft* 11 no. 4 (1985): 22–24.
12. Wolfgang van den Daele, "Der Fötus als Subjekt und die Autonomie der Frau," in *Frauensituation*, eds. Uta Gerhardt and Yvonne Schütze (Frankfurt, 1988), 189–215; Brigitte Jordan and Susan L. Irwin, "The Ultimate Failure: Court-Ordered Caesarean Section," in *New Approaches to Human Reproduction: Social and Ethical Dimensions*, eds. Linda M. Whiteford and Marilyn L. Poland (Boulder and London: Westview Press, 1989), 13–24; Marilyn L. Poland, "Ethical Issues in the Delivery of Quality Care to Pregnant Indigent Women," in *New Approaches to Human Reproduction:*

Social and Ethical Dimensions, eds. Linda M. Whiteford and Marilyn L. Poland (Boulder and London: Westview Press, 1989), 42–50; and *Newsweek* (September 1990).

13. Laura Nsiah-Jefferson and Elaine J. Hall, "Reproductive Technology," in *Healing Technology*, ed. Catherine Strother Radcliffe (Ann Arbor, 1989), 104.

14. Wolfgang van den Daele, "Technische Dynamik, *Soziale Welt* 2/3 (1983): 161.

15. Hans Jonas, *Technik, Medizin und Ethik* (Frankfurt, 1985), 22; underlined in the original.

16. Ibid., 19 f.

17. Gerhard Amendt, *Der neue Klapperstorch* (Herbstein, 1986), 27 (underlined by Elisabeth Beck-Gernsheim).

18. Horst Spielmann and Richard Vogel. "Gegenwärtiger Stand und wissenschaftliche Probleme bei der In-vitro-Fertilisierung des Menschen," in *Der codierte Leib*, eds. Alexander Schuller and Nikolaus Heim (Zürich, 1989), 27; and Urban Wiesing, "Ethik, Erfolg und Ehrlichkeit," *Ethik in der Medizin* 1 (1989): 66–82.

19. "Current *in vitro* insemination research is concentrating on ending infertility, but in the future it will probably be more concerned with early diagnosis and the prevention of genetic anomalies," according to Anne McLaren at the Fifth World Congress on In Vitro Insemination and Embryo Transfer in Norfolk, United States, 1987, quoted from Renate D. Klein, Das Geschäft mit der Hoffnung (Berlin, 1989), 253.

20. Kurt Semm, Director of the Women's Clinic in Kiel and head of the Kiel embryo-transfer team, in a lecture at the Hermann Ehlers Academy in Kiel on 5 March 1985. Quoted in, Anita Idel, "Natur und Technik, "Documentation of the Congress 19–21 April 1986, Bonn (Cologne, 1986), 620.

21. Arthur Levin quoted in Gena Corea, *MutterMaschine* (Berlin, 1986b), 124.

22. Wolfgang Friederich, "Samenbanken, Hintertür der Sterilisation," *Pro familia magazin:* 3 (1985) 22–24.

23. Gena Corea, "Die Zukunft unserer Welt," Documentation of the Congress 19–21 April 1986, Bonn (Cologne, 1986a), 24.

24. Wolfgang van der Daele, "Technische Dynamik," *Soziale Welt* 2/3 (1986): 152.

25. Hans Jonas, *Technik, Medizin und Ethik* (Frankfurt, 1985), 44 and 51.

26. Rita Arditti, Renate Duelli Klein, and Shelley Minden, Introduciton in *Retortenmütter*, eds. Rita Arditti, Renate Duelli Klein, and Shelley Minden (Reinbek, 1985), 12.

27. Saul Rosenzweig and Stuart Adelman, "Parental Determination of the Sex of Offspring," *Journal of Biological Sciences* 8 (1976): 335–346.

28. Kathy Keeton, *Woman of Tomorrow* (New York, 1985), 213.

29. Gena Corea, *MutterMaschine* (Berlin, 1986b) 150 ff.

30. Bentley Glass in his farewell address as president of the influential American lobby AAAS, quoted from Reinhard Löw, *Leben aus dem Labor* (Munich, 1985), 179.

31. Cornelia Helfferich, "Mich wird es schon nicht erwischen," in *Bauchlandungen*, eds. Monika Häussler et al. (Munich, 1983), 94.

32. Elisabeth Beck-Gernsheim. *Die Kinderfrage* (Munich, 1988).
33. Hermann van Laer, "Die demographische Entwicklung in der Bundesrepublik—Eine Bestandsaufnahme," *Sozialwissenschaftliche Informationen* 15 (1986): 60.
34. Monika Häussler, "Von der Enthaltsamkeit zur verantwortungsbewußten Fortpflanzung," in *Bauchlandungen*, eds. Monika Häussler et al. (Munich, 1983), 65.
35. Barbara Katz Rothman, "Die freie Entscheidung und ihre engen Grenzen," in *Retortenmütter*, eds. Rita Arditti et al. (Reinbek, 1985), 23 (underlined in the original).
36. Ibid., 23–25.
37. Angela Davis, *Women, Race, and Class* (New York: Random House, 1983), 210.
38. Naomi Pfeffer and Anne Wollett. *The Experience of Infertility* (London, 1983).
39. Personal accounts from ibid., 38.
40. Hans Harald Bräutigam and Liselotte Mettler, *Die programmierte Vererbung* (Hamburg, 1985), 54–86, supplemented by Spielmann and Vogel, "Gegenwärtiger Stand," in *Der codierte Leib*, eds. Alexander Schuller and Nikolaus Heim (Zurich, 1989), 39.
41. Summarized from Horst Spielmann and Richard Vogel, "Gegenwärtiger Stand," in *Der codierte Leib*, eds. Alexander Schuller and Nikolaus Heim (Zurich, 1989); and Urban Wiesing, "Ethik, Erfolg und Ehrlichkeit," *Ethik in der Medizin* 1 (1989).
42. Hans Harald Bräutigam and Liselotte Mettler, *Die programmierte Vererbung*, (Hamburg, 1985), 65.
43. Naomi Pfeffer and Anne Woolett, *The Experience of Infertility* (London, 1983).
44. Jacques Testart, *Das transparente Ei* (Munich, 1988), 18–20.
45. Barbara Katz Rothman, "Die freie Entscheidung und ihre engen Grenzen," in *Retortenmütter*, eds. Rita Arditti et al. (Reinbek, 1985), 28.
46. Ibid., 29.
47. Hans Harald Bräutigam and Liselotte Mettler, *Die programmierte Vererbung* (Hamburg, 1985), 52.
48. Kurt Semm, quoted from Idel, "Natur und Technik," 63.
49. Quoted by Alice Solomon, "Integrating Infertility Crisis Counselling into Feminist Practice," *Reproductive and Genetic Engineering* 1 (1988): 43.

4 From the Wish for a Child to the Planned Child

1. This chapter is a reworked and expanded version of the essay "Lifestyles of the Future," published in Jens Joachim Hesse, et al., *Zukunftswissen und Bildungsperspektiven* (Baden-Baden, 1988).
2. This historical development is recounted in detail in Elisabeth Beck-Gernsheim, *Die Kinderfrage* (Munich, 1988) and *Mutterwerden* (Frankfurt, 1989).
3. Andreas Flitner, *Konrad, sprach die Frau Mama . . .* (Berlin, 1982), 21.

4. Ibid.
5. Franz X. Kaufmann et al. "Familienentwicklung in Nordrhein-Westfalen," IBS-Materialien 17, University of Bielefeld 1984, 10.
6. *McCall's* (July 1984): 126.
7. Hanna Papanek, "Family Status Production: The 'Work' and 'Non-Work' of Women," *Signs* 4 no. 4 (Summer 1979): 775–781.
8. Joan Beck, *How to Raise a Brighter Child* (Glasgow, 1979), 1.
9. Philippe Aries, *Geschichte der Kindheit* (Munich, 1978).
10. Franz X. Kaufmann, et al. "Familienentwicklung—generatives Verhalten im familialen Kontext," *Zeitschrift für Bevölkerungswissenschaft* 4 (1982): 530.
11. Hartmut von Hentig, Foreword, in *Geschichte der Kindheit*, Philippe Aries (Munich, 1978), 34.
12. See, e.g., Statistisches Bundesamt ed. *Von den zwanziger zu den achtziger Jahren* (Stuttgart, 1987), 22.
13. Hans Harald Bräutigam, and Liselotte Mettler, *Die programmierte Vererbung* (Hamburg, 1985).
14. Wolfgang van den Daele, *Mensch nach Maß?* (Munich, 1985a).
15. Quoted in Claudia Roth, "Hundert Jahre Eugenik: Gebärmutter im Fadenkreuz," in *Genzeit* (Reinbek, 1985), 100 f.
16. Reinhard Löw, *Leben aus dem Labor* (Munich, 1975), 179.
17. See, e.g., Hans Jonas, *Technik, Medizin und Ethik* (Frankfurt, 1985); and Wolfgang van den Daele, *Mensch nach Maß?* (Munich, 1985a).
18. Wolfgang Friederich, "Samenbanken. Hintertür der Sterilisation," *pro familia magazin* 3 (1985): 22–24.
19. Jean Renvoize, *Going Solo: Single Mothers by Choice* (London, 1985).
20. Wolfgang van den Daele, "Technische Dynamik und gesellschaftliche Moral," *Soziale Welt* 2/3 (1986): 157.
21. Interview with Jeremy Rifkin in *Natur*, 9 (1987): 54.
22. Bundesminister der Justiz, *Der Umgang mit dem Leben*, ed. Bundesminister der Justiz (Bonn, 1987), 34.
23. Gena Corea, *MutterMaschine* (Berlin, 1986b), 297.
24. Wolfgang Michal, "Der (un)heimliche Erguß," *Geo Wissen* 1 (May 1989): 3.
25. Quoted from Gena Corea, "The Reproductive Brothel," in *Man-Made Women. How New Reproductive Technologies Affect Women*, eds. Gena Corea, et al. (London, 1985), 44.
26. Wolfgang van den Daele, *Mensch nach Maß?* (Munich, 1985a), 141.
27. Jürgen Neffe, "Mein Bauch gehört dir. Leihmutter," *Geo Wissen* 1 (May 1989): 112.
28. Gena Corea, "Industrialisierung der Reproduktion," in *Frauen gegen Gen- und Reproduktionstechnologien*, eds. Paula Bradish, et al. (Munich, 1989): 67.
29. Wolfgang van der Daele, *Mensch nach Maß?* (Munich, 1985a) 141.
30. See, e.g., Gena Corea, "Industrialisierung der Reproduktion," in *Frauen gegen Gen- und Reproduktionstechnologien*, eds. Paula Bradish, et al. (Munich, 1989) 67; and *Der Spiegel* 15 (1987): 253.
31. Quoted from Gena Corea, *MutterMaschine* (Berlin, 1986), 24.

32. Joan Rothschild, "Engineering Birth: Toward the Perfectibility of Man?" in *Science, Technology, and Social Progress*, ed. Steven L. Goldman (Bethlehem, PA, 1988).
33. Traute Schroeder-Kurth, "Pränatale Diagnostik," *Geistige Behinderung* 3(1988a): 180–89.
34. Jacques Testart, *Das transparente Ei* (Munich, 1988), 23 f.
35. Wolfgang von den Daele, "Eugenik im Angebot," *Begabung und Erziehung* (May 1985b): 41–54.
36. Jeremy Rifkin, *Kritik der reinen Unvernunft* (Reinbek, 1987), 77 f.
37. Wolfgang von den Daele, "Eugenik im Angebot," *Begabung und Erziehung* (May 1985b): 51 f.
38. Joan Rothschild, *Engineering Birth* (Bethlehem, PA, 1988).
39. Jeremy Rifkin, *Kritik der reinen Unvernunft* (Reinbek, 1987).
40. See the section "Uncontrolled Application, or, Medicine as Subpolitics."
41. See Bernd Guggenberger in *Frankfurter Allgemeine Zeitung* (March 1989).
42. Ulrich Beck, *Risikogesellschaft* (Frankfurt, 1986); Ulrich Beck, *Gegengifte* (Frankfurt, 1988); and Wolfgang van den Daele, "Politische Steuerung, faule Kompromisse, Zivilisationskritik," *Forum Wissenschaft* 1 (1987): 40–45.
43. Jacques Testart, *Das transparente Ei* (Munich, 1988).
44. Manfred Stauber, "Psychosomatische Aspekte der homologen und heterologen Insemination." Lecture given at the 9th Convention for Further Education in Applied Sexual Medicine, 13–17 June 1985 in Heidelberg.
45. Peter Petersen, "Sondervotum zum abschließenden Bericht der Arbeitsgruppe In-vitro-Fertilisation . . . ," in *Der Bundesminister fur Forschung und Technologie* (Munich, 1985), 56.

5 Pathways of Normalization and Constructions of Acceptance

1. Bernd Guggenberger in *Frankfurter Allgemeine Zeitung* (4 March 1989).
2. Peter Petersen, "Psychosomatik und die vatikanische Instruktion," in *Lebensbeginn und Menschenwürde*, ed. Stephan Wehowsky (Frankfurt, 1987), 140.
3. Susanne Davies-Osterkamp, "Sterilität als Krankheit?" *Wege zum Menschen* 2 (February–March 1990): 44–56.
4. Ulrich Beck, *Risikogesellschaft* (Frankfurt, 1986), 295.
5. Horst Spielmann and Richard Vogel, "Gegenwärtiger Stand und wissenschaftliche Probleme bei der In-vitro-Fertilisierung des Menschen," in *Der codierte Leib*, eds. Alexander Schuller and Nikolaus Heim (Zürich, 1989), 53.
6. Jacques Testart, *Das transparente Ei* (Munich, 1988), 37.
7. Ulrich Beck, *Gegengifte* (Frankfurt, 1988), 52.
8. Wolfgang van den Daele, "Kulturelle Bedingungen der Technikkontrolle durch regulative Politik," in *Technik als sozialer Prozeß*, ed. Peter Weingart (Frankfurt, 1989a), 205.
9. Konrad Adam in *Frankfurter Allgemeine Zeitung* (3 November 1989).
10. Wolfgang van den Daele, "Technische Dynamik und gesellschaftliche Moral," *Soziale Welt* 2/3 (1986): 149–72.

11. Ernst Winnacker in *Frankfurter Allgemeine Zeitung* (February 1989).
12. Hans Wilhelm Michelmann and Liselotte Mettler, "Die In-vitro-Fertilisation als Substitutionstherapie," *Lebensbeginn und Menschenwürde* ed. Stephan Wehowsky (Frankfurt, 1987), 43.
13. Erich Saling, one of the leading German gynecologists, quoted in Eva Schindele, *Gläserne Gebär-Mütter*, (Frankfurt, 1990).
14. Wolfgang van den Daele, "Kulturelle Bedingungen der Technikkontrolle durch regulative Politik," in *Technik als sozialer Prozeß* ed. Peter Weingart (Frankfurt, 1989a) 208.
15. Ulrich Beck, *Gegengifte* (Frankfurt, 1988) 41.
16. Hans Wilhelm Michelmann and Liselotte Mettler, "Die In-vitro-Fertilisation als Substitutionstherapie," *Lebenbeginn und Menschenwürde* ed. Stephan Wehowsky (Frankfurt, 1987), 43.
17. Wolfgang van den Daele, "Kulturelle Bedingungen der Technikkontrolle durch regulative Politik," in *Technik als sozialer Prozeß*, ed. Peter Weingart (Frankfurt, 1989a), 212.
18. Ibid., 212 f.
19. Ivan Illich, "Entmündigende Expertenherrschaft," *Entmündigung durch Experten*, ed. Ivan Illich (Reinbek, 1979), 14.
20. Arthur Levin, quoted from Gena Corea, *MutterMaschine* (Berlin, 1986), 124.
21. Carl Wood and Peter Singe, quoted in Gena Corea and Cynthia de Wit "Current Developments and Issues," *Reproductive and Genetic Engineering* 2 (1989): no. 2, 162.
22. Hans Harald Bräutigam and Liselotte Mettler, *Die programmierte Vererbung* (Hamburg, 1985) 135 f.
23. "According to various studies the success rate spoken of is between 10 and 20%, but these figures as a rule refer to transferred embryos with a subsequent pregnancy. Upon closer scrutiny, the success rate which is crucial to the women, i.e. the birth of a child, is considerably lower . . . The discrepancy between the purported success rates and the number of children that are actually born results from phrasing these success rates in terms of the various stages of work. Fertilizable ova are not found in all women, and insemination in the test tube and embryo transfer does not succeed in all cases. Moreover, only 50 to 70% of pregnancies result in the birth of a child. One reason for this is the number of so-called hormonal pregnancies, which are identifiable only through a brief increase in the HCG hormone and lead to a normal menstruation. Whether these are real "pregnancies" is a matter of controversy . . . Another reason is that the number of ectopic pregnancies and miscarriages, which is just over 40%, is higher than the percentage in normal pregnancies." Urban Wiesing, "Ethik, Erfolg und Ehrlichkeit" *Ethik in der Medizin* 1(1989): 69 f. See also Gena Corea, "What the King Cannot See," *Embryos, Ethics and Women's Rights*, ed. Elaine Baruch et al. (New York, 1988), 90. f.
24. Urban Wiesing, "Ethik, Erfolg und Ehrlichkeit," *Ethik in der Medizin* 1 (1989): 72.
25. Renate D. Klein, ed. *Das Geschäft mit der Hoffnung* (Berlin, 1989).
26. Rolf P. Bach, *Babyhandel mit der Dritten Welt* (Reinbek, 1986), 28 f.

27. Peter Weingart, "Großtechnische Systeme," in *Technik als sozialer Prozeß*, ed. Peter Weingart (Frankfurt, 1989b), 189 f.
28. See Chapter 2.
29. Ernst Winnacker in *Frankfurter Allgemeine Zeitung* (8 February 1989).
30. Hans Harald Bräutigam and Liselotte Mettler, *Die programmierte Vererbung* (Hamburg, 1985), 139.
31. Christoph Lau, "Risikodiskurse," *Soziale Welt* 3 (1989): 419.
32. Ibid., 433.
33. Hubert Markl, *Genetik und Ethik* (Stuttgart, 1989), 41.
34. Ibid., 41.
35. Quoted from *Frankfurter Rundschau* (March 1990).
36. Hans Harald Bräutigam and Liselotte Mettler, *Die programmierte Vererbung* (Hamburg, 1985), 142–44.
37. Quoted in Gena Corea, "Die Zukunft unserer Welt," *Frauen gegen Gentechnik und Reproduktionstechnik* (Cologne, 1986a), 25 and Gena Corea, "Industrialisierung der Reproduktion" in *Frauen gegen Gen- und Reproduktionstechnologien*, eds. Paula Bradish et al. (Munich, 1989), 63.
38. "I sometimes ask myself: why, in fact, shouldn't a married couple be able to donate a four-celled entity so that bone-marrow cells can be cultivated from it? For these could be used to heal leukemia," Interview with Liselotte Mettler in *Der Spiegel* 3 (1986): 171.
39. Monika Lans-Zumstein in Rechtsausschuß-Sekratariat des Deutschen Bundestags, *Zusammenstellung der Stellungnahmen . . .* (Bonn, 1990), 98.
40. Werner Schmid in *Neue Zürcher Zeitung* (20 January 1988).
41. Quoted from Wolfgang van de Daele, "Technische Dynamik und gesellschaftliche Moral," *Soziale Welt* 2/3 (1986): 156.
42. Hubert Markl, *Genetik und Ethik* (Stuutgart, 1989), 35.
43. Rainer Hohlfeld, "Die zweite Schöpfung . . . ," in *Der codierte Leib*, eds. Alexander Schuller and Nikolaus Heim (Zurich, 1989), 240.
44. Dr. Popvic, Executive Director of the General Medical Council of the State of Hessen, in a public hearing of the SPD in the State Parliament of Hessen in Gaby Zipfel, "Fortlanzungsmedizin," in *Mamma*, ed. Kristine von Soden (Berlin, 1989), 152.
45. The Ethics Report of the American Fertility Society in Gena Corea, "Industrialisierung der Reproduktion," in *Frauen gegen Gen- und Reproduktionstechnologien*, eds. Paula Bradish, et al. (Munich, 1989), 68.
46. Michael Joyce quoted in *Die Zeit* 19 (May 1990): 18.
47. Horst Spielmann and Richard Vogel, "Gegenwärtiger Stand und wissenschaftliche Probleme . . . ," in *Der codierte Leib*, eds. Alexander Schuller and Nikolaus Heim (Zurich, 1989), 43.
48. Hectographed working paper for the Working Group on Genetic Research (Bonn, 1989), 2.
49. Mathias Greffrath in *Die Zeit* (October 1989).
50. Konrad Adam in *Frankfurter Allgemeine Zeitung* (8 February 1989).
51. Ulrich Beck, *Risikogesellschaft* (Frankfurt, 1986), 82.
52. Ibid., 372 f.
53. Ibid., 372.

6 Brave New Health

1. Ulrich Mergner, et. al., "Gesundheit und Interesse," *Psychosozial* 2 (1990): 18.
2. Johann Jürgen Rohde, *Soziologie des Krankenhauses* (Stuttgart, 1974), 130.
3. Barbara Katz Rothman, *The Tentative Pregnancy* (London, 1988).
4. Werner Schmid in *Neue Zürcher Zeitung* (January 1988).
5. Traute Schroeder-Kurth, "Pränatale Diagnostik," *Geistige Behinderung* 3 (1988a): 182.
6. Ibid., 187.
7. Jörg Schmidtke in *Süddeutsche Zeitung* (9 August 1990).
8. Kai Kranen, "Chorea Huntington," in *Medizinische Genetik in der Bundesrepublik Deutschland*, ed. Traute M. Schroeder-Kurth (Frankfurt, 1989), 69 and 71.
9. This case history has been taken, for the most part, word-for-word from Jens Asendorpf, "Die Brisanz der Genomanalyse," in *Denkanstöße '90*, ed. Heidi Bohnet-von der Thüsen (Munich, 1989), 79–82.
10. A series of advertisements for the (association) German Chemical Industry printed in large newspapers (e.g. *Süddeutsche Zeitung*) (August 1990).
11. Traute Schroeder-Kurth, "Indikationen für die genetische Familienberatung," *Ethik in der Medizin* 1 (1989b): 202.
12. Ibid.
13. Werner Fuhrmann, "Genetische Beratung aus der Sicht eines Humangenetikers," in *Medizinische Gentechnik in der Bundesrepublik Deutschland*, ed. Traute M. Schroeder-Kurth (Frankfurt, 1989), 14.
14. Jörg Schmidtke in *Süddeutsche Zeitung* (9 August 1990).
15. Wolfgang van den Daele, *Mensch nach Maß* (Munich, 1985a).
16. Wolfgang van den Daele, "Das zähe Leben des präventiven Zwangs," in *Der codierte Leib*, eds. Alexander Schuller and Nikolaus Heim (Zurich, 1989b), 222 f.
17. Ulrich Beck, *Gegengifte* (Frankfurt, 1988), 57.
18. Wolfgang van den Daele, "Das zähe Leben des präventiven Zwangs," in *Der codierte Leib*, eds. Alexander Schuller and Nikolaus Heim (Zurich, 1989b), 211.
19. Joan Furman-Seaborg, *The Foetus as Patient, the Woman as Incubator* (Dublin, 1987).
20. Wolfgang van den Daele, "Der Foetus als Subjekt und die Autonomie der Frau," in *Frauensituation*, eds. Uta Gerhardt and Yvonne Schütze (Frankfurt, 1988), 207.
21. Ulrich Beck, *Risikogesellschaft* (Frankfurt, 1986), 216 f.
22. Wolfgang van den Daele, "Das zähe Leben des präventiven Zwangs," in *Der codierte Leib*, eds. Alexander Schuller and Nikolaus Heim (Zurich, 1989b), 208.
23. Hans Harald Bräutigam and Liselotte Mettler, *Die programmierte Vererbung* (Hamburg, 1985), 138.
24. Eva Schindele, *Gläserne Gebär-Mutter* (Frankfurt, 1990), 64.
25. Ibid.

26. Ibid., 65 f.
27. Ibid., 66.
28. Dietrich Berg, *Schwangerschaftsberatung und Perinatologie* (Stuttgart, 1988), 45.
29. Christa Fonatsch in an interview in *Bild der Wissenschaft* 7 (1986): 56.
30. Bundesministerium für Jugend, Familie und Gesundheit, *Genetische Beratung* (Bonn, 1979).
31. Walther Vogel in Maria Reif, *Frühe Pränataldiagnostik und genetische Beratung* (Stuttgart, 1990) in foreword.
32. Rayna Rapp, "Chromosomes and Communication," *Medical Anthropology Quarterly* 2 (1988): 154.
33. Dietrich Ritschl, "Die Unschärfe ethischer Kriterien," in *Medizinische Genetik in der Bundesrepublik Deutschland*, ed. Traute M. Schroeder-Kurth (Frankfurt, 1989), 136. On the problems of non-directive counseling, see also Wolfgang van den Daele, "Technische Dynamik und gesellschaftliche Moral," *Soziale Welt* 2/3 (1986): 155 f.; Wolfgang van den Daele, "Das zähe Leben des präventiven Zwangs," in *Der codierte Leib*, eds. Alexander Schuller and Nikolaus Heim (Zurich, 1989b), 221; Barbara Katz Rothman, "Die freie Entscheidung und ihre engen Grenzen," in *Retortenmütter*, eds. Rita Arditti et al. (Reinbek, 1985), 40 ff.
34. For this case outline, which I have in large part quoted directly, I thank Traute Schroeder-Kurth (from a letter of 11 August 1987).
35. Benno Müller-Hill, "Was sollten die Gentechniker aus der Geschichte der Humangenetik lernen?" *Biologie heute* 370 (November/December 1989): 8. Similarly, James Watson writes: "We used to believe that our fate lies in the stars. Now we know that to a large extent it lies in our genes." Quote in *Time* (March 1989), 65.
36. Julian Huxley quoted in Jeremy Rifkin, *Kritik der reinen Unvernunft* (Reinbek, 1987), 85.
37. See Irmgard Nippert, *Die Geburt eines behinderten Kindes* (Stuttgart, 1988).
38. Ivan Illich, *Der Nemesis der Medizin* (Reinbek, 1981).
39. Vargas Llosa, *Gegen Wind und Wetter* (Frankfurt, 1988), 25.
40. Ernst Benda in his inaugural lecture at the University of Freiburg 1984. Quoted from *Süddeutsche Zeitung* (May 1989).

Bibliography

Amendt, Gerhard. *Der neue Klapperstorch. Über künstliche Befruchtung, Samenspender, Leihmütter, Retortenzeugung* (The New Stork: On Artificial Insemination, Sperm Donors, Surrogate Mothers, In Vitro Conception). Herbstein: März Verlag, 1986.

Arditti, Rita; Renate Duelli Klein; and Shelley Minden. Introduction. In: Arditti, Klein, and Minden (eds.): *Retortenmütter. Frauen in den Labors der Menschenzüchter* (Test-Tube Mothers: Women in the Laboratories of the Human-Being Breeders). Reinbek: Rowohlt Verlag, 1985.

Ariès, Philippe. *Geschichte der Kindheit* (History of Childhood). Munich: Deutscher Taschenbuch Verlag, 1978.

Asendorpf, Jens. "Die Brisanz der Genomanalyse" (The Explosive Nature of Genome Analysis). In: Heidi Bohnet-von der Thüsen (ed.): *Denkanstöße '90. Ein Lesebuch aus Philosophie, Natur-und Humanwissenschaften* (Food for Thought '90: A Reader of Philosophy, Natural Sciences, and Social Sciences). Munich: Piper Verlag, 1989.

Bach, Rolf P. *Gekaufte Kinder. Babyhandel mit der Dritten Welt* (Bought Children: The Baby Trade with the Third World). Reinbek: Rowohlt Verlag, 1986.

Beck, Joan. *How to Raise a Brighter Child*. Glasgow: Fontana Books, 1979.

Beck, Ulrich. *Risikogesellschaft. Auf dem Weg in eine andere Moderne* (Risk Society: Toward a New Modernity). Frankfurt: Suhrkamp Verlag, 1986.

———. *Gegengifte. Die organisierte Unverantwortlichkeit* (Antidotes: The Organized Lack of Responsibility). Frankfurt: Suhrkamp Verlag, 1988.

———, and Wolfgang Bonss: "Verwissenschaftlichung ohne Aufklärung?" (Scientizing Without Educating?). In: Beck and Bonss (eds.): *Weder Sozialtechnologie noch Aufklärung? Analysen zur Verwendung sozialwissenschaftlichen Wissens* (Neither Social Technology Nor Education? Analyses of the Use of Knowledge in the Social Sciences). Frankfurt: Suhrkamp Verlag, 1989.

———; Michael Brater; and Hansjürgen Daheim. *Soziologie der Arbeit und der Berufe. Grundlagen, Problemfelder, Forschungsergebnisse* (The Sociology of Work and the Professions: Fundamental Principles, Problem Areas, Research Results). Reinbek: Rowohlt Verlag, 1980.

Beck-Gernsheim, Elisabeth. *Die Kinderfrage. Frauen zwischen Kinderwunsch und Unabhängigkeit* (The Question of Children: Women Between the Wish for Children and Independence). Munich: Beck Verlag, 1988.

———. *Mutterwerden—der Sprung in ein anderes Leben* (Becoming a Mother: The Leap into Another Life). Frankfurt: Fischer Verlag, 1989.

Berg, Dietrich. *Schwangerschaftsberatung und Perinatologie* (Pregnancy Counseling and Perinatology). Stuttgart: 1988.

Bräutigam, Hans Harald, and Liselotte Mettler. *Die programmierte Vererbung. Möglichkeiten und Gefahren der Gentechnologie* (Programmed Heredity: The

Opportunities and Dangers of Genetic Technology). Hamburg: Hoffmann und Campe, 1985.

Bundesminister der Justiz (Federal Minister of Justice) (ed.). *Der Umgang mit dem Leben. Fortpflanzungsmedizin und Recht* (Dealing with Human Life: Reproductive Medicine and the Law). Bonn: 1987.

Bundesministerium für Jugend, Familie und Gesundheit (Federal Ministry for Youth, the Family, and Health) (ed.). *Genetische Beratung. Ein Modellversuch der Bundesregierung in Frankfurt und Marburg* (Genetic Counseling: an Experiment in Frankfurt and Marburg). Bonn: 1979.

Corea, Gena. "The Reproductive Brothel." In: Gena Corea et al.: *Man-Made Women. How New Reproductive Technologies Affect Women*. London: Hutchinson, 1985.

——. "Die Zukunft unserer Welt" (The Future of Our World). In: *Frauen gegen Gentechnik und Reproduktionstechnik* (Women Against Genetic Technology and Reproductive Technology). Documentation of the Congress 19–21 April 1985 in Bonn. Cologne: Kölner Volksblatt Verlag, 1986a.

——. *MutterMaschine. Reproduktionstechnologien—von der künstlichen Befruchtung zur künstlichen Gebärmutter* (MotherMachine: Reproductive Technologies from Artificial Insemination to the Artificial Womb). Berlin: Rotbuch Verlag, 1986b.

——. "What the King Can Not See." In: Elaine Hoffman Baruch et al. (eds.). *Embryos, Ethics, and Women's Rights. Exploring the New Reproductive Technologies*. New York: Harrington Park Press, 1988.

——. "Industrialisierung der Reproduction" (The Industrialization of Reproduction). In: Paula Bradish et al. (eds.). *Frauen gegen Gen- und Reproductionstechnologien. Beiträge vom 2. bundesweiten Kongreß* (Women Against Genetic and Reproductive Technologies. Papers from the Second National Congress) Frankfurt, 28–30 October 1988. Munich: Verlag Frauenoffensive, 1989.

——, and Cynthia de Wit. "Current Developments and Issues: A Summary." In: *Reproductive and Genetic Engineering*. Vol. 2, No. 2, 1989.

Daele, Wolfgang van den. *Mensch nach Maß? Ethische Probleme der Genmanipulation und Gentherapie* (The Tailor-Made Human Being? Ethical Problems of Genetic Manipulation and Genetic Therapy). Munich: Beck Verlag, 1985a.

——. "Eugenik im Angebot" (Cut-Rate Eugenics). In: *Kursbuch No. 80, Begabung und Erziehung* (Aptitude and Education) (May 1985b): 41–54.

——. "Technische Dynamik und gesellschaftliche Moral. Zur soziologischen Bedeutung der Gentechnologie" (The Dynamics of Technology and Social Morality: On the Sociological Significance of Genetic Technology). In: *Soziale Welt* (Social World), No. 2/3 (1986): 149–72.

——. "Politische Steuerung, faule Kompromisse, Zivilisationskritik. Zu den Funktionen der Enquetekommission 'Gentechnologie' des Deutschen Bundestages" (Political Manipulation, Cowardly Compromises, Criticism of Civilization: On the Functions of the German Bundestag's Investigative Commission on Genetic Technology). In: *Forum Wissenschaft* (Science Forum), No. 1 (1987): 40–45.

————. "Der Fötus als Subjekt und die Autonomie der Frau. Wissenschaftlich-technische Optionen und soziale Kontrollen in der Schwangerschaft" (The Foetus as a Subject and Women's Autonomy: Scientific and Technical Options and Social Controls During Pregnancy). In: Uta Gerhardt and Yvonne Schütze (eds.): *Frauensituation. Veränderungen in den letzten zwanzig Jahren* (The Situation of Women: Changes During the Past Twenty Years). Frankfurt: Suhrkamp Verlag, 1988.

————. "Kulturelle Bedingungen der Technikkontrolle durch regulative Politik" (Cultural Preconditions for Monitoring Technology Through Regulatory Policies). In: Peter Weingart (ed.): *Technik als sozialer Prozeß* (Technology as a Social Process). Frankfurt: Suhrkamp Verlag, 1989a.

————. "Das zähe Leben des präventiven Zwangs" (The Stubborn Survival of Preventive Pressure). In: Alexander Schuller and Nikolaus Heim (eds.): *Der codierte Leib. Zur Zukunft der genetischen Vergangenheit* (The Coded Body: On the Future of the Genetic Past). Zürich: Artemis Verlag, 1989b.

Davies-Osterkamp, Susanne. "Sterilität als Krankheit?" (Is Sterility an Illness?). In: *Wege zum Menschen* (Paths Toward the Human Being). Issue 2 (February March 1990): 49–56.

Davis, Angela. *Women, Race, and Class.* New York: Random House, 1983.

Flitner, Andreas. *Konrad, sprach die Frau Mama . . . Über Erziehung und Nicht-Erziehung* (Konrad, Said His Madam Mother . . . On Child-Rearing and Nonrearing). Berlin: Severin und Siedler, 1982.

Frevert, Ute. "'Fürsorgliche Belagerung': Hygienebewegung und Arbeiterfrauen im 19. und frühen 20. Jahrhundert" ("Detained for Their Own Good": The Hygiene Movement and Working-Class Women in the Nineteenth and Early Twentieth Centuries). In: *Geschichte und Gesellschaft* (History and Society), 11th year No. 4 (1985): 420–46.

Friederich, Wolfgang. "Samenbanken. Hintertür der Sterilisation" (Sperm Banks: The Back Door of Sterilization). In: *pro familia magazin*, No. 3 (1985): 22–24.

Fuhrmann, Werner. "Genetische Beratung aus der Sicht eines Humangenetikers" (Genetic Counseling from the Perspective of a Human Geneticist). In: Traute M. Schroeder-Kurth (ed.): *Medizinische Gentechnik in der Bundesrepublik Deutschland* (Medical Genetic Technology in the Federal Republic of Germany). Frankfurt: Schweitzer Verlag, 1989.

Furman-Seaborg, Joan. *The Foetus as Patient, the Woman as Incubator.* Lecture at the Third International Interdisciplinary Congress on Women, Dublin, July 1987 (hectographed manuscript).

Greffrath, Mathias. "Der lange Arm von Chromosom Nr. 7. Visionen und Ambitionen der Top-Genforscher" (The Long Arm of Chromosome No. 7: Visions and Aspirations of the Top Researchers in Genetics). In: *TransAtlantik*, No. 12 (December 1990): 14–28.

Häussler, Monika. "Von der Enthaltsamkeit zur verantwortungsbewußten Fortpflanzung. Über den unaufhaltsamen Aufstieg der Empfängnisverhütung und seine Folgen" (From Abstinence to Responsible Conception: On the Inexorable Rise of Contraception and Its Consequences). In: Monika Häussler et al.: *Bauchlandungen. Abtreibung—Sexualität—Kinderwunsch* (Belly Flops:

Abortion—Sexuality—The Wish for a Child). Munich: Frauenbuch Verlag, 1983.

Helfferich, Cornelia. "'Mich wird es schon nicht erwischen.' Risikoverhalten und magisches Denken bei der Verhütung" ("It Can't Happen to Me": Risk-Taking and Magical Thinking in Contraception). In: Monika Häussler et al.: Bauchlandungen. Abtreibung—Sexualität—Kinderwunsch (Belly Flops: Abortion—Sexuality—The Wish for a Child). Munich: Frauenbuch Verlag, 1983.

Hentig, Hartmut von. "Foreword". In: Philippe Ariès: Geschichte der Kindheit (History of Childhood). Munich: Deutscher Taschenbuch Verlag, 1978.

Hesse, Jens Joachim et al. (eds.). Zukunftswissen und Bildungsperspektiven (Knowledge of the Future and Perspectives for Education). Baden-Baden: Nomos Verlag, 1988.

Hohlfeld, Rainer. "Die zweite Schöpfung des Menschen—eine Kritik der Idee der biochemischen Verbesserung des Menschen" (The Second Creation of Man—a Criticism of the Idea of Improving the Human Being through Biochemistry). In: Alexander Schuller and Nikolaus Heim (ed.): Der codierte Leib. Zur Zukunft der genetischen Vergangenheit (The Coded Body: On the Future of the Genetic Past). Zürich: Artemis Verlag, 1989.

Idel, Anita. "Natur und Technik" (Nature and Technology). In: Frauen gegen Gentechnik und Reproduktionstechnik (Women Against Genetic Technology and Reproductive Technology). Documentation of the Congress 19–21 April 1985 in Bonn. Cologne: Kölner Volksblatt Verlag, 1986.

Illich, Ivan. "Entmündigende Expertenherrschaft" (The Disempowering Rule of the Experts). In: Ivan Illich et al.: Entmündigung durch Experten. Zur Kritik der Dienstleistungsberufe (Disempowerment by the Experts: Toward a Critique of the Service Professions). Reinbek: 1979.

———. Die Nemesis der Medizin. Von den Grenzen des Gesundheitswesens (The Nemesis of Medicine: On the Limits of the Health System). Reinbek: Rowohlt Verlag, 1981.

Jonas, Hans. Technik, Medizin und Ethik. Zur Praxis des Prinzips Verantwortung (Technology, Medicine, and Ethics: On the Application of the Principle of Responsibility). Frankfurt: Insel Verlag, 1985.

Jordan, Brigitte, and Susan L. Irwin. "The Ultimate Failure: Court-Ordered Caesarean Section." In: Linda M. Whiteford and Marilyn L. Poland (eds.): New Approaches to Human Reproduction: Social and Ethical Dimensions. Boulder and London: Westview Press, 1989.

Jürgensen, Ortrun. "Gedanken zur manipulierten Fruchtbarkeit" (Thoughts on Manipulated Fertility). In: Wege zum Menschen (Roads Toward the Human Being), No. 2 (February–March 1990): 56–62.

Kaufmann, Franz X., et al. "Familienentwicklung–generatives Verhalten im familialen Kontext" (Family Development: Reproductive Behavior in the Context of the Family). In: Zeitschrift für Bevölkerungswissenschaft (Journal of Population Science), No. 4 (1982): 523–45.

———, et al. "Familienentwicklung in Nordrhein-Westfalen" (Family Development in North Rhine-Westphalia). IBS-Materialien (IBS Reference Papers), No. 17, University of Bielefeld 1984.

Keeton, Kathy. Woman of Tomorrow. New York: St. Martin's Press, 1985.

Klein, Renate D. (ed.). *Das Geschäft mit der Hoffnung. Erfahrungen mit der Fortpflanzungsmedizin. Frauen berichten* (The Hope Business: Experiences with Reproductive Medicine—Women Tell Their Stories). Berlin: Orlanda Frauen Verlag, 1989.

Kollek, Regine et al. (eds.). *Die ungeklärten Gefahrenpotentiale der Gentechnologie* (The Unexplored Potential Dangers of Genetic Technology). Munich: Schweitzer Verlag, 1986.

Kranen, Kai. "Chorea Huntington. Das Recht auf Wissen versus das Recht auf Nicht-Wissen" (Huntington's Chorea: The Right to Know Versus the Right Not to Know). In: Traute M. Schroeder-Kurth (ed.): *Medizinische Genetik in der Bundesrepublik Deutschland* (Medical Genetics in the Federal Republic of Germany). Frankfurt: Schweitzer Verlag, 1989.

Laer, Hermann van. "Die demographische Entwicklung in der Bundesrepublik— Eine Bestandsaufnahme" (Demographic Developments in the Federal Republic of Germany—Taking Stock). In: *Sozialwissenschaftliche Informationen* (Information from the Social Sciences), 15th year No. 1 (1986): 58–60.

Lau, Christoph. "Risikodiskurse. Gesellschaftliche Auseinandersetzungen um die Definition von Risiken" (Discourses on Risks: Social Controversies Concerning the Definition of Risks). In: *Soziale Welt* (Social World), No. 3 (1989): 418–36.

Löw, Reinhard. *Leben aus dem Labor. Gentechnologie und Verantwortung— Biologie und Moral* (Life from the Laboratory: Genetic Technology and Responsibility—Biology and Morality). Munich: Bertelsmann Verlag, 1975.

Lüscher, Kurt et al. (eds.) *Die "postmoderne" Familie. Familiale Strategien und Familienpolitik in einer Übergangszeit* (The Postmodern Family: Family Strategies and Family Policy in a Time of Transition). Constance: Universitätsverlag Konstanz, 1988.

Markl, Hubert. *Genetik und Ethik. Rede anläßlich der Verleihung des Arthur-Burkhardt-Preises* (Genetics and Ethics. A speech delivered at the award ceremony for the Arthur Burkhardt Prize) 1989, Stuttgart, 26 April 1989 (hectographed manuscript).

Mergner, Ulrich, et al. "Gesundheit und Interesse. Zur Fremdbestimmung von Selbstbestimmung im Umgang mit Gesundheit" (Health and Interests: On Outsiders' Determination of Self-Determination in Matters of Health). In: *Psychosozial*, No. 2 (1990): 7–29. Theme: Gesundheit als gesellschaftlicher Zwang (Health as a Social Compulsion).

Mettler-Meibom, Barbara. "Mit High-Tech zurück in eine autoritäre politische Kultur?" (High Tech: Regression to an Authoritarian Political Culture?) In: *Essener Hochschulblätter. Ausgewählte Reden im Studienjahr* (Notes from the Technical College of Essen: Selected Lectures from the College Year) 1988/1989. Essen 1990.

Michal, Wolfgang. "Der (un)heimliche Erguß. Samenspende und-übertragung" (The Secret Shot: Sperm Donation and Transfer). In: *Geo Wissen* (Geo Knowledge), No. 1 (May 1989). Special Issue: Sex—Geburt—Genetik (Sex—Birth— Genetics), 102 f.

Michelmann, Hans Wilhelm, and Liselotte Mettler. "Die In-vitro-Fertilisation als Substitutionstherapie" (In Vitro Insemination as Substitution Therapy).

In: Stephan Wehowsky (ed.): *Lebensbeginn und Menschenwürde* (Human Dignity and the Beginning of Life). Frankfurt: Schweitzer Verlag, 1987.

Müller-Hill, Benno. "Was sollten die Gentechniker aus der Geschichte der Humangenetik lernen?" (What Should Genetic Technologists Be Learning from the History of Human Genetics?). In: *Biologie heute* (Biology Today), No. 370 (November/December 1989): 6–8.

Neffe, Jürgen. "Mein Bauch gehört dir. Leihmutter" (My Belly Belongs to You: The Surrogate Mother). In: *Geo Wissen* (Geo Knowledge), No. 1 (May 1989), Special Issue: Sex—Geburt—Genetik (Sex—Birth—Genetics), 104–12.

Nippert, Irmgard. *Die Geburt eines behinderten Kindes. Belastung und Bewältigung aus der Sicht betroffener Mütter und ihrer Familien* (The Birth of a Handicapped Child: The Burden and Coping with It, as Seen by Affected Mothers and Their Families). Stuttgart: Enke Verlag, 1988.

Nsiah-Jefferson, Laura, and Elaine J. Hall. "Reproductive Technology: Perspectives and Implications for Low-Income Women and Women of Color." In: Kathryn Strother Ratcliff (ed.): *Healing Technology: Feminist Perspectives*. Ann Arbor: University of Michigan Press, 1989.

Papanek, Hanna. "Family Status Production: The 'Work' and 'Non-Work' of Women." In: *Signs*, 4th year/No. 4 (Summer 1979): 775–81.

Petersen, Peter. "Sondervotum zum abschließenden Bericht der Arbeitsgruppe In-vitro-Fertilisation, Genomanalyse und Gentherapie" (Special Vote on the Final Report of the Working Group on In Vitro Insemination, Genome Analysis, and Genetic Therapy). In: Der Bundesminister für Forschung und Technologie (Federal Minister for Research and Technology) (ed.): *In-vitro-Fertilisation, Genomanalyse und Gentherapie. Bericht der gemeinsamen Arbeitsgruppe des Bundesministers für Forschung und Technologie und des Bundesministers der Justiz* (In Vitro Insemination, Genome Analysis, and Genetic Therapy. Report of the Joint Working Group of the Federal Minister for Research and Technology and the Federal Minister of Justice). Munich: Schweitzer Verlag, 1985.

———. "Psychosomatik und die vatikanische Instruktion" (Psychosomatics and the Vatican's Instructions). In: Stephan Wehowsky (ed.): *Lebensbeginn und Menschenwürde* (Human Dignity and the Beginning of Life). Frankfurt: Schweitzer Verlag, 1987.

Pfeffer, Naomi, and Anne Woollett. *The Experience of Infertility*. London: Virago Press, 1983.

Poland, Marilyn L. "Ethical Issues in the Delivery of Quality Care to Pregnant Indigent Women." In: Linda M. Whiteford and Marilyn L. Poland (eds.): *New Approaches to Human Reproduction: Social and Ethical Dimensions*. Boulder and London: Westview Press, 1989.

Postman, Neil. *Die Verweigerung der Hörigkeit* (The Refusal to Be Enslaved). Frankfurt: Fischer Verlag, 1988.

Rapp, Rayna. "Chromosomes and Communication: The Discourse of Genetic Counseling." In: *Medical Anthropology Quarterly*, Vol. 2 (1988): 143–57.

Rechtsausschuß-Sekretariat des Deutschen Bundestags (Secretariat of the Judiciary Committee of the German Bundestag) (ed.). *Zusammenstellung der Stellungnahmen zur öffentlichen Anhörung am 9. März 1990 zum*

Embryonenschutzgesetz (Survey of Statements at the Public Hearing on 9 March 1990 on the Embryo Protection Law). Bonn 1990 (hectographed manuscript).

Reif, Maria. *Frühe Pränataldiagnostik und genetische Beratung. Psychosoziale und ethische Gesichtspunkte* (Early Prenatal Diagnostics and Genetic Counseling: Psychosocial and Ethical Perspectives). Stuttgart: Enke Verlag, 1990.

Renvoize, Jean. *Going Solo: Single Mothers by Choice.* London: Routledge & Kegan, 1985.

Rifkin, Jeremy. *Kritik der reinen Unvernunft* (Critique of Pure Unreason). Reinbek: Rowohlt Verlag, 1987.

Ritschl, Dietrich. "Die Unschärfe ethischer Kriterien. Zur Suche nach Handlungsmaximen in genetischer Beratung und Reproduktionsmedizin" (The Vagueness of Ethical Criteria: On the Search for Guidelines for Behavior in Genetic Counseling and Reproductive Medicine). In: Traute M. Schroeder-Kurth (ed.): *Medizinische Genetik in der Bundesrepublik Deutschland* (Medical Genetics in the Federal Republic of Germany). Frankfurt: Schweitzer Verlag, 1989.

Rohde, Johann Jürgen. *Soziologie des Krankenhauses* (The Sociology of the Hospital). Stuttgart: Enke Verlag, 1974.

Rosenzweig, Saul, and Stuart Adelman. "Parental Determination of the Sex of Offspring: The Attitudes of Young Married Couples with University Education." *Journal of Biosocial Sciences*, 8th year (1976): 335–46.

Roth, Claudia. "Hundert Jahre Eugenik: Gebärmutter im Fadenkreuz" (A Century of Eugenics: The Uterus as a Target). In: Claudia Roth (ed.): *Genzeit. Die Industrialisierung von Pflanze, Tier und Mensch. Ermittlungen in der Schweiz* (The Age of the Gene: The Industrialization of Plants, Animals, and Human Beings. Investigations in Switzerland). Zürich: Limmat Verlag, 1987.

Rothman, Barbara Katz. "Die freie Entscheidung und ihre engen Grenzen" (Free Choice and Its Narrow Limits). In: Rita Arditti et al. (eds.): *Retortenmütter. Frauen in den Labors der Menschenzüchter* (Test-Tube Mothers: Women in the Laboratories of the Human-Being Breeders). Reinbek: Rowohlt Verlag, 1985.

——. *The Tentative Pregnancy: Prenatal Diagnosis and the Future of Motherhood.* London: Pandora Press, 1988. German edition: Schwangerschaft auf Abruf. Marburg 1989.

Rothschild, Joan. Engineering Birth: "Toward the Perfectibility of Man?" In: Steven L. Goldman (ed.): *Science, Technology, and Social Progress.* Lehigh University Press, 1988.

Schindele, Eva. Gläserne Gebär-Mutter. *Vorgeburtliche Diagnostik—Fluch oder Segen* (The Transparent Uterus: Prenatal Diagnostics—Curse or Blessing?). Frankfurt: Fischer Verlag, 1990.

Schroeder-Kurth, Traute. "Pränatale Diagnostik. Probleme der Indikationsstellung und zukünftige Trends" (Prenatal Diagnostics: Problems in Establishing Indications and Future Trends). In: *Geistige Behinderung* (Mental Handicaps), No. 3 (1988a): 180–89.

——. "Vorgeburtliche Diagnostik" (Prenatal Diagnostics). In: Traute Schroeder-

Kurth and Stephan Wehowsky (eds.): *Das manipulierte Schicksal. Künstliche Befruchtung, Embryotransfer und pränatale Diagnostik* (Manipulating Fate: Artificial Insemination, Embryo Transfer, and Prenatal Diagnostics). Frankfurt: Schweitzer Verlag, 1988b.

────── (ed.). *Medizinische Genetik in der Bundesrepublik Deutschland* (Medical Genetics in the Federal Republic of Germany). Frankfurt: Schweitzer Verlag, 1989a.

──────. "Indikationen für die genetische Familienberatung" (Indications for Genetic Family Counseling). In: *Ethik in der Medizin* (Ethics in Medicine), No. 1 (1989b).

Solomon, Alice. "Integrating Infertility Crisis Counselling into Feminist Practice." In: *Reproductive and Genetic Engineering*, 1st year/No. 1 (1988): 41–49.

Spielmann, Horst, and Richard Vogel. "Gegenwärtiger Stand und wissenschaftliche Probleme bei der In-vitro-Fertilisierung des Menschen" (The Current Situation and Scientific Problems in the In Vitro Insemination of Human Beings). In: Alexander Schuller and Nikolaus Heim (eds.): *Der codierte Leib. Zur Zukunft der genetischen Vergangenheit* (The Coded Body: On the Future of the Genetic Past). Zürich: Artemis Verlag, 1989.

Ständige Deputation des Deutschen Juristentages (Standing Deputation of the Council of German Jurists' Congress) (ed.). *Verhandlungen des 56. Deutschen Juristentages* (Proceedings of the 56th German Jurists' Congress, Berlin: 1986). Munich: 1986.

Statistisches Bundesamt (Federal Bureau of Statistics) (ed.). *Von den zwanziger zu den achtziger Jahren. Ein Vergleich der Lebensverhältnisse der Menschen* (From the Twenties to the Eighties: A Comparison of People's Living Conditions). Stuttgart: Kohlhammer Verlag, 1987.

Stauber, Manfred. *Psychosomatische Aspekte der homologen und heterologen Insemination.* Vortrag auf den 9. Fortbildungstagen für praktische Sexualmedizin (Psychosomatic Aspects of Homologous and Heterologous Insemination. Lecture given at the 9th Convention for Further Education in Applied Sexual Medicine), Heidelberg, 13–17 June 1985 (hectographed manuscript).

Testart, Jacques. *Das transparente Ei* (The Transparent Ovum). Munich: Schweitzer Verlag, 1988.

Vargas Llosa, Mario. *Gegen Wind und Wetter* (Against the Wind and Weather). Frankfurt: Suhrkamp Verlag, 1988.

Weingart, Peter et al. *Rasse, Blut und Gene. Geschichte der Eugenik und Rassenhygiene in Deutschland* (Race, Blood, and Genes: The History of Eugenics and Racial Hygiene in Germany). Frankfurt: Suhrkamp Verlag, 1988.

────── (ed.). *Technik als sozialer Prozeß* (Technology as a Social Process). Frankfurt: Suhrkamp Verlag, 1989a.

──────. "'Großtechnische Systeme'—ein Paradigma der Verknüpfung von Technikentwicklung und sozialem Wandel?" ("Large-Scale Technical Systems"—A Paradigm of the Relation Between Technological Development and Social Change?). In: Peter Weingart (ed.): *Technik als sozialer Prozeß* (Technology as a Social Process). Frankfurt: Suhrkamp Verlag, 1989.

Wiesing, Urban. "Ethik, Erfolg und Ehrlichkeit. Zur Problematik der In-vitro-

Fertilisation" (Ethics, Success, and Honesty: On the Problems Related to In Vitro Insemination). In: *Ethik in der Medizin* (Ethics in Medicine), No. 1 (1989): 66–82.

Zipfel, Gaby. "Fortpflanzungsmedizin. Die künstliche Befruchtung weiblicher Identität" (Reproductive Medicine: The Artificial Insemination of Female Identity). In: Kristine von Soden (ed.): *Mamma*. Berlin: Elefanten Press, 1989.

Index